The environment has become exposed to a range of chemical contaminants from a wide variety of sources, including the application of pesticides. Regulation of chemical accumulation in the environment has frequently been hampered by difficulties in cooperation between disparate disciplines in the natural, social and political sciences.

This volume forms the conclusion of five years' collaboration between toxicologists, economists and lawyers in the understanding and analysis of the problem of accumulative chemicals. As well as a case study of the accumulation of pesticides in groundwater in one particular region (the European Union), the book forms a general study of the value of interdisciplinary approaches in environmental policy-making.

This volume will be a valuable resource for a broad group of academics and researchers in the area of environmental science and environmental policy. It will also form a useful supplementary reference text for courses in environmental policy, science, economics and toxicology.

TIMOTHY SWANSON is a Professor in the School of Public Policy, University College London, and a Programme Director at the Centre for Social and Economic Research on the Global Environment.

MARCO VIGHI is Professor of Applied Ecology at the University of Milan, and a member of the Scientific Advisory Committee on Ecotoxicity and Environment of the European Union (CSTEE).

Regulating chemical accumulation in the environment

The integration of toxicology
and economics in environmental
policy-making

Edited by Timothy Swanson
and Marco Vighi

CAMBRIDGE
UNIVERSITY PRESS

CAMBRIDGE UNIVERSITY PRESS
Cambridge, New York, Melbourne, Madrid, Cape Town, Singapore, São Paulo, Delhi

Cambridge University Press
The Edinburgh Building, Cambridge CB2 8RU, UK

Published in the United States of America by Cambridge University Press, New York

www.cambridge.org
Information on this title: www.cambridge.org/9780521593106

First published 1998
This digitally printed version 2008

A catalogue record for this publication is available from the British Library

Library of Congress Cataloguing in Publication data
Regulating chemical accumulation in the environment : the integration
of toxicology and economics in environmental policy making / edited
by Timothy Swanson and Marco Vighi.
    p.   cm.
ISBN 0 521 59310 7 (hardbound)
1. Environmental policy.   2. Environmental chemistry – Economic
aspects.   3. Environmental toxicology.   I. Swanson, Timothy M.
II. Vighi, M.
GE170.R445   1998
363.738–dc21   98-13354   CIP

ISBN 978-0-521-59310-6 hardback
ISBN 978-0-521-08856-5 paperback

For Richard Lloyd, a valued colleague
and calming influence.

It was a pleasure to work with you.

Contents

Contributors

Michael G. Faure
METRO, University of Maastricht, PO Box 616,
6200 MD Maastricht, The Netherlands

Paolo Ferrario
Istituto di Ingegneria Agraria, Via Celoria 2,
20133 Milan, Italy

Magnus Johannesson
Stockholm School of Economics, Box 6501,
S-113 83 Stockholm, Sweden

Per-Olov Johansson
Stockholm School of Economics, Box 6501,
S-113 83 Stockholm, Sweden

Jürgen G. J. Lefevere
METRO, University of Maastricht, PO Box 616,
6200 MD Maastricht, The Netherlands

Richard Lloyd[†]
Fisheries Department, Ministry of Agriculture,
Food and Fisheries, Burnham-on-Crouch, UK

Robin Mason
Centre for Communications Systems Research and
Department of Applied Economics, Sidgwick Avenue,
Cambridge CB3 9DE, UK

Jane Press
Fondazione Eni Enrico Mattei, Corso Magenta 63,
20123 Milan, Italy

Carolina Sbriscia Fioretti
Dupont de Nemours Italiana, Via Volta 16,
20093 Cologno Monzese, Milan, Italy

† Deceased

Tore Söderqvist
Department of Economics, Stockholm School of Economics,
Royal Swedish Academy of Sciences, Box 50005,
S-10450 Stockholm, Sweden

Timothy Swanson
School of Public Policy, University College London,
Gower Street, London WC1E 6BT, UK

Marco Vighi
Department of Environmental Sciences, University of Milan,
Via Emmanueli 15, 20126 Milan, Italy

Giuseppe Zanin
Centro di Studio sulla Biolgia ed il Controllo delle
Piante Infestanti, CNR – Agripolis, 35020 Legnaro,
Padua, Italy

Preface: The regulation of pesticides in Europe – past, present and future

Marco Vighi, Carolina Sbriscia Fioretti and Timothy Swanson

Moving towards a preventive approach

Toxicology and ecotoxicology are disciplines that have developed in response to a need for information about the possible damages that might result from chemical usage. During the seventies a shift occurred from *a posteriori* control of chemical impacts to the prevention of this type of damage. The change in emphasis occurred first in the scientific community and then in the administrative and political spheres. As a result, many important regulations were approved for application across Europe. The essence of these regulations was to require preliminary information on the toxicology and ecotoxicology of chemicals in order to make available data needed for a preventive risk assessment of the characteristics of the marketed chemicals.

In particular, the Toxic Substances Control Act (US EPA, 1978) in the USA and the Sixth Amendment to the Directive on Dangerous Substances (EEC Council Directive, 1979) in Europe require the development of a basic set of information before a new chemical substance may be marketed. The required data set dealt with several characteristics of the substance (chemical structure, use patterns, physico-chemical properties, analytical methods, etc.) and includes toxicological and ecotoxicological tests at different levels of complexity in relation to the amount of the substance produced and the results at the preliminary levels (see Table 1).

The challenge to the scientific community was therefore: to what extent can the impacts of the chemicals be predicted by reference to this relatively limited set of data? The complexity of the question depends mainly on the fact that any kind of evaluation for a potentially harmful substance must take into account two types of factors – the first intrinsic to the substance, the second related to the extrinsic conditions (environmental factors,

Table 1. *Toxicological and ecotoxicological tests at three different complexity levels required by Directive 79/831/EEC (sixth amendment of Directive 67/548/EEC on the approximation of the laws, regulations and administrative provisions relating to the classification, packaging and labelling of dangerous substances)*

	Base set	Level 1	Level 2

Toxicological tests	Acute toxicity
	  Oral
	  Inhalation
	  Cutaneous
	  Skin irritation
	  Eye irritation
	  Skin sensitisation
	
	Subacute toxicity
	  NOEL at 28 days
	
	Other effects
	  Mutagenicity
		Fertility study
		Teratology study
		Subchronic and/or chronic toxicity study
		Additional mutagenisis studies
			Chronic toxicity study
			Carcinogenicity study
			Fertility study
			Teratology study
			Acute and subacute toxicity study on a second species
			Additional toxicokinetic studies

Ecotoxicological tests	Effects on organisms
	  Acute toxicity for fish
	  Acute toxicity for *Daphnia magna*
	
	Degradation
	  Biotic
	  Abiotic
		Algal growth inhibition test
		Prolonged toxicity study with *Daphnia magna*
		Test on a higher plant
		Test on an earthworm
		Prolonged toxicity study with fish
		Test for species accumulation
			Additional tests for accumulation, degradation and mobility
			Prolonged toxicity study with fish (including reproduction)
			Additional toxicity study (acute and subacute) with birds
			Additional toxicity study with other organisms
			Absorption/desorption study

Note:
NOEL, no observed effect level

population exposure, etc.) and their interactions. Obviously, a preliminary report based upon laboratory data can take into account, at most, only the intrinsic properties of the substance.

This toxicological and ecotoxicological risk assessment of new chemicals is a key feature of Directive 91/414/EEC concerning the placing of plant protection products on the market (EEC Council Directive, 1991). In brief, it requires that the applicant for the authorisation of a plant protection product produce information on the uses, efficacy, and chemical and toxicological properties of the compound. The dossier must be submitted to a commission of experts of the Member States of the European Union (EU) and a final monograph must be drawn up containing a complete evaluation of the information provided. In particular, for the ecotoxicological evaluation, in addition to the results of a wide set of toxicological tests (see Table 2), information must be provided on the distribution and ultimate fate of the chemical in the major environmental compartments (soil, water, air).

The risk assessment of the chemical must be based on the evaluation of toxicology–exposure ratios (TERs) calculated as the ratios of the predicted environmental concentrations (PECs) and various toxicological end points (e.g. the dose needed to kill 50% of a sample population of experimental animals (LD_{50}) and the highest dose that does not produce any evidence of an effect (NOEL, no observed effect level) on a number of terrestrial and aquatic living organisms. At the current time, there is an active debate occurring at European level regarding the standardisation of the criteria for evaluation of this dossier and for the drawing up of the monographs. Special care must be taken in the selection of methods applicable to the range of various agronomic and environmental conditions typical of the European territory and capable of producing comparable results in the complex situation of the different Member States of the EU. The heterogeneity of the European land mass makes prediction of overall outcomes a very complex undertaking.

The predictive approaches of toxicology and ecotoxicology have been developed (see Vighi *et al.*, Chapter 4, this volume) in an attempt to provide suitable answers to these difficult questions, but the role of prediction is necessarily a limited one. All chemical substances have the capacity for some environmental impacts. In addition, many of the most useful chemical substances will necessarily have some capacity to accumulate within the enviroment. This is because toxicity (at least for the target organisms) is

Table 2. *Toxicological and ecotoxicological tests required by Directive 91/414/EEC, for placing plant protection products on the market*

Toxicological tests	Ecotoxicological tests
Acute toxicity	Effect on birds
Oral	Oral acute toxicity
Inhalation	Short-term dietary toxicity
Cutaneous	Effects on reproduction
Intraperitoneal	
Skin irritation	Effects on aquatic organism
Eye irritation	Acute toxicity to fish
Skin sensitisation	Chronic toxicity to fish
	Effects on reproduction and
Subacute toxicity	growth of fish
Subacute oral toxicity (28 days)	Bioaccumulation in fish
90 days' feed trials	Acute toxicity for *Daphnia*
Additional exposure routes	*magna*
	Fertility test for *Daphnia magna*
Chronic toxicity	Effects on algal growth
Long term oral toxicity and	
carcinogenicity	Effects on other non target
Mutagenicity	organisms
	Acute toxicity for honeybees
Effects on reproduction	and other beneficial
Teratology studies	arthropods
Multigenerational studies on mammals	Toxicity for earthworms and
	other non-target soil
Studies on mammals metabolism	macroinvertebrates
Adsorption, distribution and excretion	Effects on non-target soil
patterns	microorganisms
Metabolic patterns	Effects on other non-target
	organisms at risk
Studies on neurotoxicity	Effects on biological methods
	for treatment of waste water
Additional studies	
Effects of metabolites	
Studies on mode of action	
Effects on cattle and domestic animals	
Medical and epidemiological data	

an important characteristic of agricultural chemicals, and accumulation in the environment results from chemical stability (i.e. non-degradation) in the general environment, also a desirable trait of commercial chemicals. Clearly, the capacity to predict chemical toxicity and accumulation is not sufficient in itself for adequate chemical regulations; prediction must be combined with a measure that determines when toxicity and accumulativeness reach 'undesirable' levels. This will depend upon the meaning given to 'undesirable levels of accumulation', and it must also depend upon the various conditions under which a chemical is used.

The setting of quality objectives and standards for pesticides

In Chapter 4 of this book, Vighi *et al.* describe the procedures adopted by various international organisations for the setting of quality objectives, in particular for the protection of the aquatic environment. All of the approaches are extremely laborious and require an amount of toxicological information which is available for only a relatively small number of potentially dangerous compounds. As a consequence, the number of scientifically sound quality objectives produced and accepted by internationally acknowledged organisations is very low (no more than a few hundred) in comparison with the huge number of potential contaminants.

In particular, the Scientific Advisory Committee on Ecotoxicity and Environment (CSTEE) of the EU has produced Water Quality Objectives for about 100 substances, selected and published in a list of priority chemicals (CSTEE/EEC, 1994a). Among them, 32 are pesticides and the figures proposed are reported in Table 3. Pesticides figure prominently in these objectives because they are specifically designed to be biocides, toxic substances particularly effective against some target groups of living organisms (plants, insects, etc.). The Quality Objectives are aimed at protecting the whole ecosystem, including the most sensitive species of the natural biological communities. It may be observed, however, in many cases these Water Quality Objectives are extremely low, sometimes orders of magnitude below the guidelines for drinking water. It should be stressed that a Water Quality Objective is not a legal standard but only a scientific suggestion. It may be used as an indicator of the need for suitable interventions for the protection of the natural environment, at local or national level, but it is a function of ecological as well as political and economic factors.

Table 3. *Water Quality Objectives (WQO) for pesticides proposed by the EU/CSTE*

Compounds	WQO (mg/m³)	Compounds	WQO (mg/m³)
Atrazine	1.00	Linuron	1.00
Azinphos-ethyl	0.01	Malathion	0.01
Azinphos-methyl	0.01	Methylparathion	0.01
Biphenyl	1	Mevinphos	0.01
Carbon tetrachloride	10.00	Omethoate	0.01
Chlorophenylid	0.10	Parathion	0.01
DDT	0.002	Pentachlorophenol	1.00
Demeton-methyl	0.10	Pyrazon	0.10
1,3-Dichloropropene	10.00	Simazine	1.00
Dichlorovos	0.001	Sulcofenuron	10.00
Endosulfan	0.001	2, 4, 5-T	1.00
Fenitrothion	0.01	Tributyltin oxide	0.001
Fenthion	0.01	Trifluralin	0.10
Flucofenuron	0.10	Triphenyltin acetate	0.01
Hexachlorozene HCB	0.01	Triphenyltin chloride	0.01
Hexachlorocyclohexane	0.01	Triphenyltin hydroxide	0.01

The zero tolerance approach for pesticides in drinking water

A completely different approach is followed by the EU for the management of xenobiotics in drinking water and, in particular, of pesticides. The EU has applied a policy of zero tolerance toward the presence of pesticides in drinking water since 1980. At that time it adopted a policy establishing the maximum acceptable level of pesticides in drinking water at the concentration of 0.1 μg/l in Directive 80/778/EEC (EEC Council Directive, 1980). This is taken as a '*practically zero*' level of tolerance, considering the analytical detection limit for most pesticides at the time of promulgation of the Directive.

The philosophy of the EU in establishing this zero tolerance standard is based on the following principles.

(1) Xenobiotics are substances not present in nature before the era of synthetic chemicals and, in particular, pesticides are toxic substances by definition. Ideally they should not be present in

the natural environment. In practice, all possible measures of preventive management and of control of emission must be applied in order to maintain the level of these substances 'as low as possible', especially in particularly valuable environmental resources.

(2) The contamination level in drinking water must be more strictly controlled than in food. Usually, every day a person imbibes about two litres of drinking water originating from the same source, and this on a lifelong basis. Thus, in the case of the presence of pesticides, even if at a toxicologically safe level, there is a continuous exposure to the same potentially toxic agent. On the other hand, every day a person may eat different agricultural products that come from various origins and do not contain the same pesticide residues. In this case the possible exposure (again to toxicologically safe levels) is occasional and discontinuous.

This position is often criticized as *arbitrary* and *non-scientific*. It must be emphasised, however, that the methodology for establishing any standard must always contain some level of arbitrariness. For example, the procedure for establishing the acceptable daily intake (ADI) standards, used as the basis for the toxicologically sound World Health Organizaton (WHO) Guidelines, applies a number of safety factors, which contain a large degree of arbitrariness and cannot be considered as a rigorously scientific procedure. Therefore, when compared to the WHO Guidelines, for example, the EU limit is not 'toxicologically incorrect' but 'philosophically different'. Standard-making must always take into account a number of potentially incommensurable factors.

What has been the outcome of the zero tolerance policy?

Despite the use of this policy over a period of almost 30 years, there is none the less a problem of pesticide contamination in groundwater across Europe. In many European countries the concentration levels of specific pesticides (e.g. atrazine) have breached the EU set standard.

What have been the responses to the already existing accumulation of pesticides and the predictable future accumulation of pesticides in the agricultural regions of Europe? First, local derogations to the directive have been allowed up to the concentrations believed to be toxicologically safe (e.g. according to WHO Guidelines) and for the time needed to undertake suitable control measures. Secondly, the local governments have often banned the offending chemical, disallowing its further sale or application in the regions where it has already accumulated. Finally, the EU has now revised its standard, in order to allow further accumulation of pesticides in groundwater. This is the result of the abandonment of the so-called 'cocktail standard', which proscribed aggregate accumulation of all chemicals in drinking water. None of these responses is geared to correcting the underlying problems by using the regulatory approach.

Therefore, irrespective of the desirability of the EU zero tolerance approach, its implementation has clearly been problematic. The proper implementation of the basic objectives of the EU regulatory strategy must be carefully considered, so that the important environmental objectives may be attained.

Outputs: how should chemical accumulation in drinking water be regulated?

According to the opinion of the CSTEE, and taking into account that about two-thirds of the drinking water of the EU comes directly from natural groundwater and consumed without treatment, drinking water should be regulated taking into account three different points of view (CSTEE(1), EEC 1994*b*):

(1) The *ethical and quality-oriented* point of view, which may be summed up as the widely held preference for non-polluted and pristine water sources.
(2) The *technological* point of view, which considers the possibility of controlling the use of chemical substances to avoid their presence in drinking water.
(3) The *scientific* point of view, dealing with the following:
  Consumer health protection: the concentration of substances in drinking water should be such that any consumer can drink the water for a lifetime without risk of adverse health effects.
  Resource protection: measures should be taken so that in future water resources will not be at risk of possible pollution.

The core of this approach is to prevent all unnecessary and unwanted chemical contamination of drinking water. There are many objections, however, to the EU approach, which are based mainly on its potentially high costs (of preventive control measures and their effects on the competitiveness of agricultural products). This indicates that it is important to achieve the correct balance of environmental and agricultural objectives in this area. The regulation of drinking-water quality necessarily involves a balancing of two important but potentially conflicting societal goals: agricultural production and environmental quality.

One of the objects of this volume is to assess this trade-off in the context of a case study concerning one particular agricultural chemical, atrazine. In the course of this case study we hope to demonstrate the methodology that might be used in the balancing of the important but inconsistent objectives.

The second object of this volume is to demonstrate how such a trade-off may be implemented. Clearly, the EU approach to implementation has not yet been successful and alternative approaches must be considered. The implementation of environmental objectives may be pursued through a combination of agronomic, environmental, economic and political means. They may be given effect through a range of different approaches:

> The regulation of further chemical usage in regions of already contaminated water sources.
> The approval of more 'environmental friendly' compounds in relation to the environmental resources to be protected.
> The management of land use and agricultural practices in relation to the protection of particularly valuable and vulnerable natural resources.

We will examine in the context of our case study this range of possible approaches, and how to use each of them most effectively. We will also indicate where we believe the EU approach went wrong, and how it might be rectified.

As a final output we hope to produce a volume that will be instructive in the understanding of how various disciplines must interrelate in the development of policies concerning the environment. Economists, toxicologists and lawyers were all necessary for this research to be undertaken, and for this volume to be its result. Appropriate policy-making in the future must develop these sorts of hybrid endeavours in order to reach the ultimate policy objectives.

xx   M. Vighi *et al.*

References

CSTEE/EEC Ecotoxicity Section (1994a). EEC water quality objectives for chemicals dangerous to aquatic environments. *Review of Environmental Contamination and Toxicology*, **137**, 83–110.

CSTEE/EEC (1994b). Opinion of the Scientific Advisory Committee concerning the revision of the Drinking Water Directive. CSTE/94/31. (Mimeo)

EEC Council Directive 79/831 (1979). *Official Journal of the European Community* L259/10.

EEC Council Directive 80/778 (1980). *Official Journal of the European Community* L229/11.

EEC Council Directive 91/414 (1991). *Official Journal of the European Community* L230/1.

USEPA (United States Environmental Protection Agency) (1978). Toxic Substances Control Act. *US Federal Register*, **43**, 4108.

Acknowledgements

This volume is the end product of five years' collaboration between toxicologists, economists and lawyers in the understanding and analysis of the problem of accumulative chemicals, and it represents a truly integrated approach to the subject. The project was funded by the European Science Foundation in order to encourage interdisciplinary cooperation in environmental policy-making and this volume bears witness to that vision. The contributors to this volume wish to acknowledge the efforts and inputs of the other members of the group who participated in the project but who did not author chapters, including Jan Henrik Kock, Michael Pugh and Lars Bergman. We are grateful for their participation in the project. Finally we would like to thank all of those individuals at Cambridge University Press who have helped to see the project through to fruition, especially Sandi Irvine who rendered a rough manuscript readable. We are grateful for all of the support we have received in the accomplishment of this project.

Introduction

1 Regulating chemical accumulation: an integrated approach

Timothy Swanson

The problem under consideration

In July 1980 the European Commission issued a Directive on drinking-water quality (80/778/EEC) setting a maximum admissible concentration for 71 distinct parameters. One of the most strictly regulated substances in the directive was the set of chemical pesticides. The European Commission adopted a 'practically zero' level of permissible contamination for these substances. The limit for any individual pesticide product was set at the trace level of 0.1 μg/l; a 'cocktail' standard for the allowed aggregate level of contamination by all chemical pesticides was set at 0.5 μg/l. These were levels of chemical contamination that were only just detectable under then-existing monitoring technologies. The Commission's standard was intended as a clear and unequivocal pronouncment against the accumulation of chemicals within the drinking water of the EEC.

Despite this pronouncement against chemical accumulation, pesticides have been accumulating in groundwater over the past 15 years to such an extent that several substances have breached the allowed concentration in groundwater in many of the agricultural districts across the European Union (EU) (see, generally, Bergman and Pugh, 1994). This is important because two-thirds of the EU citizenry continue to acquire their drinking-water supplies from untreated groundwater, i.e. directly from the aquifers underlying their communities. In adopting its tough stance against chemical accumulation, it had been the object of the European Commission to stimulate a comprehensive strategy of pesticide management (based on agricultural, land use and pesticide management). However, the continued accumulation of pesticides in European groundwater supplies placed the EU in the position of choosing between two poor options: either the relaxation of its earlier drinking-water quality directive or the costly treatment of groundwater prior to delivery to consumers. The latter option would generate additional costs estimated at around £10 per annum for each consumer of treated water supplies. (Söderqvist, 1994). The former option would entail a substantial loss of political and regulatory credibility. The Commission was caught between two equally unsavoury options.

The EU's approach to the resolution of this dilemma to date has been to do some of each. It has allowed the individual states to select the measures required to meet the directive's standards, in order to allow for cost-effective implementation based upon local conditions (Faure and Lefevre, Chapter 10, this volume). It has also relaxed the 'cocktail standard' for aggregate accumulation, in order to allow for the already observed additional accumulation of chemicals in groundwater sources.

The usual approach of the Member States to the problem of chemical accumulation has been to implement product-specific bans when a specific chemical has breached the EU standards. The disallowance of a market to a chemical found to accumulate in groundwater would seem to be a straightforward method for proscribing chemical accumulation. Once again it would seem to be intended to send a strong and clear signal (at national level) that accumulative chemicals are not to be allowed in use. Nevertheless agricultural chemical accumulation in groundwater supplies continues apace, even in those countries where such bans have already been implemented. The example of groundwater contamination in the maize-growing districts of norther Italy is a case in point (Sbriscia Fioretti *et al.*, Chapter 2, this volume).

The Po River Valley is an important maize-growing district with an aggressive weed problem. In the absence of an active weed control programme, it has been estimated that 31% to 38% of the average maize yield would be lost to weed encroachment. In the 1950s selective herbicide application became the primary mode of weed control, and in 1964 this strategy was extended to maize production in Italy, with the introduction of the chemical atrazine. Atrazine was a stunningly successful pesticide, providing very effective and reliable weed control for many seasons following its introduction. Of course chemical-induced selection implies the need for an evolving weed control programme, and atrazine required supplementation by other chemical products throughout the 1970s. This resulted in increasing volumes as well as increasing numbers of herbicides being applied to the Italian countryside throughout the seventies and into the eighties (see chapter 2, Table 2.3). The level of application of atrazine remained relatively constant throughout this period, even though it was being increasingly supplemented by other chemicals as well.

The Directive on drinking-water quality was finally implemented in Italy on 2 August 1985, and the monitoring of groundwater supplies was initiated on an official basis. As a consequence it was discovered that many of the communities within the agricultural district of the Po Valley were being provided with drinking water containing pesticides (including atrazine) in breach of the EU standard. In order to enforce the standard the relevant authorities (districts) initiated local, then district-level proscriptions on the application of atrazine. These product-specific bans were slow to begin (with 67 000 hectares regulated initially in 1987) but rose to include entire regions (367 000 hectares total) by 1990. Nevertheless these location-specific prohibitions were deemed inadequate and, in 1991, the product atrazine was banned from all sale or use within the state of Italy, both in those areas in which it had accumulated and in those in which it had not. A nation-wide ban of this nature will of course help to reduce the cost of enforcing the prohibition in those areas in which it is most needed. In addition, the perception was that the government was sending a signal to chemical producers and users that accumulative products were not to be tolerated, with the foreclosure of markets to those substances which demonstrably breached these standards. Bans on specific offending chemical products are often hoped to have such broad impacts on the incentives for the use of these and related chemicals (Toman and Palmer, 1997).

Despite the clarity of the policy stance against accumulation, both within the EC Directive and in the foreclosure of markets, there is little evidence that the rate of accumulation of such chemicals is slowing. Continued monitoring identifies wells newly in breach of the guidelines on account of past years of chemical applications; due to prevailing geological conditions, it is possible for maximum concentration levels to be achieved years, even decades, after application has ceased. *Even more alarmingly the newly marketed chemicals frequently exhibit characteristics equally as accumulative as those which they are replacing* (Sbriscia Fioretti *et al.*, Chapter 2 this volume; Mason, Chapter 8 this volume). After the proscription of specified chemical products on the grounds of their accumulative nature, many of the replacement chemicals used in their stead exhibit characteristics which will cause them to appear in groundwater in similar concentrations after an equivalent amount of time. The strong stand taken across and within the EU against chemical accumulation has been having little or no effect on the number or quality of accumulative substances being produced and applied within the Union. This is the primary reason that the EU was forced to relax its 'cocktail standard' on pesticide accumulation. The policy measures preventing the accumulation of specific chemical products are not having the effect of shaping the characteristics of their replacements sufficiently, and one chemical after another is accumulating in the groundwater.

How can it be the case that such strong policy measures have so little impact? It is the object of this volume to explain this conundrum. We hope to demonstrate both the reasons for the inefficacy of existing policies and the essence of an effective approach to regulating chemical accumulation. The remainder of this chapter provides an overview of our approach, and an indication of our conclusions. I recommend that the interested reader read each of the individual chapters to acquire the full story on chemical accumulation and its regulation.

An overview of the volume: empirical studies

Part I of the volume presents two chapters which attempt to dissemble the problematic pesticide into its constituent components. This allows the ensuing discussion to pursue the subject at a more fundamental level. It is not the chemical nature of the products that is problematic nor their widespread use per se, rather it is the specific characteristics of certain chemical products that gives rise to their accumulative nature. Part I of this volume identifies these characteristics, and sets forth an analysis that ascertains their relative contribution to a chemical's use and usefulness. This analysis will then be helpful later in the unravelling of the nature of the policy failures in this area, but initially it provides an excellent introduction to the nature of problematic chemical substances in general and of atrazine (and its substitute substances) in particular.

The first chapter in this section demonstrates both the need for agricultural chemicals and the need for a policy explicitly addressing the contamination resulting from their use (Sbriscia Fioretti *et al.*). Since the 1950s, chemical-based strategies have been the preferred form of weed control, and in their absence it has been estimated that up to a third of crop production would be lost. On the other hand, many agricultural chemicals have been designed in such a fashion as to ensure their accumulation in groundwater. This is because many of these chemicals (herbicides in particular) are designed to use the natural flow of precipitation to transport the chemical from the surface (where it is applied) into the soil. It is within the soil that the chemical then acts upon the germinating seeds and root matter of the weedy plant. In essence, the hydraulic cycle (from atmosphere to surface through soil and other living matter and back into the atmosphere via respiration and evaporation) is used as the transport vector through which the chemical may travel to make contact with the target organisms. For this reason, herbicides have been explicitly designed in order to react primarily with water rather than alternative media (i.e. the atmosphere or organic sphere).

The groundwater contamination problem arises because some of the natural flow of water leaks out of this cycle, and becomes relatively stagnant within various substrata. In these so-called 'sinks' the chemical substance accumulates under circumstances (out of contact with light, air or organic substances) in which it is difficult for further biodegradation to occur. The chemical's natural affinity for water has led it down a dead-end, where it will continue to accumulate so long as degradation and recharge rates are low. Groundwater aquifers are one of those dead-ends in which chemical substances are capable of being found.

For these reasons the two traits of a chemical that are most likely to determine its rate of accumulation within groundwater are: (1) its relative affinity for reacting with water relative to the other basic media (the organic sphere, the atmosphere), and (2) its absolute rate of reactivity or persistence. 'Affinity' is measured by virtue of partition coefficients which determine the rate at which the substance will react with alternative media when simultaneously exposed to them; for example, the K_{oc} coefficient states a chemical's relative affinity for organic carbon and water media. 'Persistence' is usually measured by the amount of time required for the loss of half of the original mass of the chemical substance through reactivity (the chemical's 'half-life'). The product of these two measures is combined into something termed the 'GUS index': a measure of a chemical's in-built propensity for accumulation within groundwater. Clearly, chemical substances with longer half-lives and higher relative affinities for water will have a greater proportion of their initial applications finding their way into groundwater sinks.

Of course both water affinity and persistence are in-built characteristics of useful chemicals. Water affinity provides the substance with its transport vector – to take it where it needs to be. Persistence reduces the need for multiple applications because it allows for the correct amount of the chemical to be on hand at the time that its action is needed. It does this by reducing the rate at which the chemical reacts with non-target substances; i.e. by reducing its general rate of reactivity or biodegradation. Hence, it is no accident that these chemicals accumulate in groundwater; the propensity for accumulation is a by-product of the same characteristics that render the chemical useful. This point is pursued further in an empirical analysis of the demand for the various characteristics of pesticides (Söderqvist, chapter 3, this volume). This study looks more closely at atrazine and its various substitute chemicals, and assesses the relative demand for the various characteristics which distinguish them from one another. The characteristics of useful chemicals examined there include:

(1) Persistence (half-lives).
(2) Reliability (GUS index for pre-emergents).
(3) Effectiveness (kill rate).
(4) Toxicity (lethal dose).
(5) Regulation (banned status).
(6) Age (years on market).

Unsurprisingly, this study demonstrates that the effectiveness (kill rate) of the chemical is the single most important facet of the substance; users are clearly willing to pay more for chemical substances which are more effective in removing the targeted organisms. There are other, more surprising, results from this study, but these will be addressed in the discussion later in this chapter concerning the policy studies regarding atrazine.

At this juncture, the importance of the studies in Part I is to demonstrate the nature and object of chemical design: it is a matter of in-built chemical characteristics related to very specific targets and objectives. The contest between crops and their competitors is an important and continuing one. Agricultural chemicals are not blunt instruments; they are carefully designed to perform specific functions along charted routes through the environment. This section of the volume demonstrates the complex nature of chemical design, and the range of characteristics across which chemical manufacturers must operate (persistence, affinity, toxicity, kill rate). The choices that manufacturers make regarding these various parameters are determined by what makes for a useful chemical substance in the context within which they are used. This implies that chemical accumulation is a linked outcome, not an unintended consequence, of chemical production and application. It is probably incorrect to view the societal objective as the prohibition of all accumulative substances (unless the entirety of the benefits of chemical applications are to be foregone), as opposed to the calibration of chemical design (and application) in order to balance the benefits of chemical usefulness against the cost of chemical accumulation. Part I of this volume details how the various traits of a chemical are demanded in agriculture, and how these same traits can contribute to various forms of unintended, but necessarily linked, consequences such as accumulation in the groundwater. It demonstrates the basic nature of the social problem of regulating the traits that cause chemical accumulation: the trade-off between groundwater purity and chemical effectiveness.

The valuation of resource contamination

Part II of this volume then launches into the problematic region of environmental valuation. In a previous volume (Bergman and Pugh, 1994), we discussed the importance of undertaking a cost–benefit analysis of the EU drinking-water standard for pesticide accumulation, in order to calibrate the cost of the EU policy concerning chemical accumulation against its benefits. In that volume we reported a rough estimate of the cost of the EU policy; as mentioned previously, the cost of removing pesticides from groundwater by the use of granular activated carbon filters in the Po Valley region was estimated to be around £10 per consumer per annum (Söderqvist, 1994). Now we turn to the task of estimating the benefits.

The benefits of removing these trace chemicals are more difficult to calibrate. On account of their extremely low levels of concentration in groundwater, it will require many years of continuous exposure before that exposure accumulates to levels which are toxicologically meaningful. The toxicological procedure for extrapolating an acceptable daily intake (ADI) for any given chemical based upon various indicators such as its acute toxicity is widely accepted and not under examination here, but a large amount of uncertainty must remain in a context such as this one. This is because toxicologists must operate in laboratory environments and on time scales much shorter than a normal human life-span; they are simply unable to replicate the conditions which are prevalent in the environment in assessing their likely impacts. The low level conditions of contamination prevailing in groundwater are hence not discernibly costly under standard toxicological measures; yet a large degree of uncertainty remains, precisely because these measures are not suited to the problem of long-term low dosage induced responses.

Part II describes the toxicological and the economic approaches to risk assessment under conditions of uncertainty. In addition to the uncertainty related to the definition of a toxicologically based ADI, the extremely low levels of the EU standard for drinking water (much lower than any estimated ADI) renders it technically impossible to estimate the risks of such levels of contamination based upon toxicologically relevant considerations. The toxicologists frankly admit that the EU standards are based on foundations other than the toxicological; they are 'philosophically different' from the World Health Organization (WHO) toxicology-based standards, relying upon ethical, technological as well as scientific precepts. The economists, on the other hand, advance the willingness-to-pay criterion for use in this region of profound uncertainty. If consumers are concerned about little-understood hazards such as low level groundwater contamination, then perhaps the best measure of the cost of these hazards is the willingness of consumers to pay to undertake efforts to avoid them. Willingness to pay (WTP) for avoidance of a risk is the preferred measure used by economists to calibrate the magnitude of these sorts of preference across individuals, and the economic analysis of this problem in Part II studies a range of different approaches to the estimation of this measure.

Part II includes two chapters presenting two distinct economic approaches to the assessment of the benefits of avoiding groundwater quality deterioration: (1) a survey of *indirect method* studies for quantifying individual responses to risk (Johannesson and Johansson, Chapter 5, this volume) and (2) a *contingent valuation survey* for the same purpose (Press and Söderqvist, Chapter 6, this volume). The survey on the indirect method studies reports on several different markets which contain risk assessments implicitly, e.g. workers accepting jobs with less risk exposure at lower wages. Empirical studies across such markets once again break down the marketed product or occupation into its constituent characteristics, and then ascertain the relative contribution of each characteristic to the differences in prices between the products. In this fashion the implicit price assessed by consumers to a characteristic such as potential hazardousness may be derived. Several studies have discussed this value with regard to the willingness to pay to avoid an incremental hazard resulting in the loss of one additional life, and found a range of estimates of between US$1 million and US$20 million. This approach is valuable when the relative risks of the alternatives are already known, as it is then possible to weight this risk according to the value that individuals are observed to place on risk avoidance. This is not so helpful in circumstances, such as this, where the risks are relatively low but uncertain.

Another indirect approach to valuing risks that is more applicable in this context is to use observed expenditures that individuals undertake in order to avoid the risk. For example, the risks and uncertainties of groundwater contamination may be avoided in part by, for example, drinking bottled water, installing a water filter or moving to an area with better water supplies. Obviously, some of these are better indicators of the willingness to pay for pure water, and all of them are actions replete with mixed motives. Nevertheless, avertive expenditures provide a market-based indicator of willingness to pay, and useful indicators of the potential value placed on the risk by individuals. Three studies regarding avertive behaviour towards groundwater contamination are reviewed in this volume, indicating WTP values regarding water contamination risks in the neighbourhood of US$1–10 per individual per week. (Johannesson and Johansson, Chapter 5, this volume).

The problem with these market-based indicators is that the willingness to pay measure should be geared as closely as possible to the actual environmental good that is being valued – in this context, pristine groundwater quality. Health risks (actual and perceived) are only one facet of this environmental good. For many centuries Europeans have been able to drink untreated groundwater piped directly from the aquifers, and then into their houses. The advent of intensive agricultural production and the introduction of chemical methods of weed control have now changed this for the first time. The continued application of large volumes of chemical pesticides will make it necessary to introduce drinking-water treatment, as is now the case in the most intensive agricultural districts, and it has denied Europeans something that was part of their natural heritage.

In addition, the loss of the pristine resource is something that the individual citizen might value for reasons other than health risks and uncertainty. There are also its effects on wildlife and other biota, general ecosystems, and general environmental degradation. For these reasons the market-based indirect methods of estimation are far too narrow. The true willingness to pay for pristine groundwater quality must allow for the inclusion of this wider range of characteristics and motivations that might be included in a willingness to pay for the underlying resource. This calls for different sorts of valuation technique.

Economists attempt to estimate the broadest range of values inherent in an environmental good by means of the construction of artificial or imaginary markets for the good. This is done by constructing an imaginary mechanism for maintaining the environmental good – in our experiment we used a groundwater management fund established through water taxation – and then a random sample of individuals are asked what tax they would be willing to pay into such a fund for the purpose of maintaining groundwater quality (Press and Söderqvist, Chapter 6, this volume). This form of study is known as a contingent valuation exercise, because it asks the individual to give a valuation of the good that is contingent upon his or her acceptance of the vehicle identified as its mode of provision (here, the groundwater management fund). This volume reports a contingent valuation study undertaken in Milan, Italy, in which the average stated willingness to pay for the maintenance of groundwater quality was about ITL640 000 (aprox. £320) per household per annum. This is a very large figure relative to average household income or average household expenditures on bottled drinking water. It indicates in part that there are many broader values at stake in the preservation of pristine resources such as groundwater than simply the narrowest measures of health risks and uncertainties.

Clearly all of these methods for risk assessment and resources valuation have their own failings and presumed biases toward underinclusiveness and overstatedness. The object of including in this volume all of these various approaches to the valuation of the risks and uncertainties associated with chemical accumulation is to demonstrate the range of methods available and the kinds of result they provide. A balanced assessment of the costs and benefits of chemicals and chemical contamination will have to consider all of these various approaches to risk assessment and environmental valuation; however, our study makes clear that individual health risk from contaminated groundwater is only one part of the overall rationale for environmental regulation. Individuals are willing to pay for regulation that takes into account the broader sets of values (wildlife, heritage, etc.) that are affected by the fact of continuing environmental degradation.

Existing market and regulatory failures

Part III of this volume concerns the failings of the existing systems of regulation regarding accumulative chemical substances. The first issue discussed is whether markets will fail to generate the correct characteristics within chemical products from the overall societal perspective. The answer to this appears to be straightforward. Intuitively it might seem that the average pesticide user would be interested in maximising water affinity and persistence, in order to enhance the overall effectiveness of the chemical. Since these two traits are related directly to the rate at which the substance will accumulate in groundwater, it would then be the case that the chemical users' (i.e. the farmers') demands for chemical characteristics would be in direct conflict with those of the groundwater users. This would be the classic form of 'externality' – where some decision-makers take choices that maximise their own objectives without regard to the implied impacts on others.

There is more than a grain of truth to this paradigm as it applies in the context of agricultural chemical usage; however, as in most instances of simple generalisation, there are many complicating factors. This is because the user's objective with respect to chemicals is not so uncomplicated as it was depicted. The user clearly does not want the chemical to exhibit the largest possible persistence, for instance. This is because land use changes over time, and desirable chemicals in one context become highly undesirable in another. Pesticide residuals are cited as a major problem of chemical usage (Sbriscia Fioretti *et al.*, Chapter 2, this volume) even by their own users, so to some extent chemical and groundwater users' objectives are aligned.

This limited appeal of persistence as a chemical characteristic is evident in the results of the hedonic analysis undertaken by Söderqvist (Chapter 3, this volume). In that study there was no clear link found between the trait of persistence and the farmers' willingness to pay for the chemical. This was one of the more surprising results from that study.

Even more surprisingly that study also found a significant but *negative* relationship between what is termed 'reliability' (estimated by the use of the GUS index in the case of pre-emergents) and the users' willingness to pay. Once again this is evidence that the conflict between chemical and groundwater users is not so clear-cut as might have been thought to be the case. This finding might have been cast off as some sort of a statistical anomaly but for the fact that there is only one other empirical study on this point, and it came to the same conclusion (Beach and Carlson, 1993). It is not at all apparent that the problem of groundwater contamination is simply the result of accumulative characteristics being chosen by chemical producers in order to satisfy the demands (for water affinity, persistence and effectiveness) of chemical users. This is part of the problem, but not the whole of it. What else can explain the in-built characteristics that contribute to the accumulation of pesticides in groundwater?

Part III describes another category of explanations for excessive accumulation as 'regulatory failures' (Mason, Chapter 8, this volume). Unlike market failures, regulatory failures do not represent conflicts between the preferences of various user groups: they are instead the strategic response by concentrated industries to various forms of regulatory structures. Sometimes a regulation, ostensibly adopted for one public purpose, may instead be turned to affect the objects of the industry it was intended to regulate. This is because an industry with one or a few firms is capable of responding to the regulator in a carefully conceived and strategic manner. Since the chemical industry is one of the world's most concentrated (in the sense that a small number of firms control a large proportion of the global market), and the markets are further subdivided through patent claims and licenses, it would not be too surprising if the industry responded strategically to proffered regulations. This means that regulations must be drafted extremely carefully in anticipation of such reactions, in order to have their intended effect.

The drinking-water Directive of 1980 and its subsequent implementation is an excellent case study to illustrate this point. This Directive (as described earlier) was intended to convey the disapproval of the community regarding all accumulative chemical pesticides – their use was not to be allowed if they were found to accumulate in groundwater above trace levels. It has already been mentioned that this is an illogical objective: it implies the prohibition of many substances precisely for those characteristics that render them useful. Unless the EU means to render entirely unlawful the use of the natural hydrological cycle as a transport vector (implying the use of natural organic and atmospheric media in its stead), then the objective as stated makes little sense.

If the EU régime is instead interpreted to imply that chemicals should be used and designed in such a manner that they are less rather than more accumulative, then this makes more sense as an objective but it remains largely unattained. The study by Mason, Chapter 8, this volume, demonstrates that the replacement chemicals for those which are specifically banned are equally as accumulative (as measured by their GUS indices) as those which they are replacing (atrazine).

How can it be that the EU régime does not generate incentives to design even marginally less accumulative chemical substances than those which have been banned? Consider the simplest case as an illustration, the case in which the industry is effectively monopolised by a patent-holding firm. How would that firm respond to the prospect of a ban being placed on its patented product should it accumulate in groundwater to the proscribed level? Would it respond by creating substitute chemicals which are less accumulative? Not necessarily. If the firm perceived itself as the most likely recipient of a patent for a substitute product, then it would perceive the threatened ban simply as a mechanism for determining when it would switch from one patented product to the other. A strategic reaction to the threat of a ban would be precisely the opposite of that which was intended. The firm would instead plan to sell quantities of the accumulating chemical so as to *ensure* that the patented chemical *did* reach the level that would engage the ban on its further sale or use. This strategy would provide the firm with the capability of choosing the time at which all further sales of the first product were disallowed. Why would it care about disallowing the future sale of this product? On the expiration of a patent, the product is then available for production by any and all firms in the industry. Banning the future use and production of the now-generic chemical makes room in the market for the newly patented product.

This is a 'pre-emption' sort of strategy. It disallows general entry into the firm's monopolised market by reason of that firm's exhaustion of an ancillary but necessary resource. Here, the EU's proscription of specific chemical products unless specified stocks of groundwater remain uncontaminated implicitly renders that groundwater supply a necessary input into chemical production and use. Once that quantity of groundwater is exhausted so is the right to manufacture the chemical. Firms with market power (e.g. current patents and the prospects for replacement patents) could respond to threatened product-specific bans by strategically exhausting the resource rather than conserving it.

Regulatory failures occur whenever regulatory mechanisms are inadequately planned and implemented within the context of market power. It is predictable that regulations will have unintended consequences when the regulated firms are not naïve in their responses. There is evidence for regulatory failure underlying these problems in the hedonic analysis by Söderqvist (Chapter 3, this volume). Two of the variables demonstrating significant relationships with market prices of chemical products are the extent of product regulation and the time that the product has been on the market; both are inversely related to the market price of the chemical. This implies that the newer, replacement chemicals are more expensive than the previous generation of chemicals (in addition to being equally as accumulative). This finding is consistent with the use of product-specific regulations as a method to phase out older increasingly inexpensive chemicals (whose patents are expiring) for replacement by the newly patented more expensive chemicals.

Part III discusses a wide range of reasons that might underlie the inefficient choice of chemical characteristics, and the inefficient accumulation of chemicals in groundwater. The most obvious explanation is the straightforward problem of externality between agricultural producers and water consumers. One group wants to use the water resource as a vector to transport its pesticides to the targets; the other wishes to consume water uncontaminated by such a use. Obviously there is a societal conflict inherent in these uses that must be taken into account in the regulation of agricultural chemicals. Less obvious is the problem that regulation itself can engender. The chemical industry is a concentrated one, and typified by producers of patented goods. These conditions are likely to give rise to relatively complicated responses to regulation. The EU's attempt to discourage chemical accumulation by banning products which accumulate is a case in point. Our empirical studies indicate that the continued production of accumulating chemicals in this context (i.e. the replacements for atrazine) is less likely to be a straightforward response to agricultural users' demands than it is a strategic reaction by producers to the product-specific regulation. If this is the case, this means that it is the form that the EU regulation has taken that has resulted in its ultimate ineffectiveness (and now the relaxation of the standard). This indicates the importance of making policies correctly in order to make them effective, and this leads us into the subject of the next part of the volume.

Optimal policies for accumulative chemicals.

Part IV outlines our suggested approach to regulating chemical accumulation. It is based on the idea of internalising the externalities inherent in the design and use of chemical characteristics, but doing so in a manner that anticipates the most obvious outlets for strategic responses.

The chemical characteristic most closely linked to chemical accumulation is persistence. A chemical that reacts slowly will persist in the same environment for a longer period of time, existing through more cycles and allowing more opportunities for its leakage into some sink. For example, a persistent chemical with an affinity for water will remain where it is applied while successive waves of the hydraulic cycle pass over it. Since there is little breakdown of the substance in the interim, it continues to flow into the hydraulic system over time, with some fraction always leaking out of the system and into a sink (such as an aquifer). Once there, its general reticence for reactivity will maintain it and allow it to accumulate. Hence the trait of persistence will always generate increased accumulation in the sinks with which it has an affinity.

Since persistence can be measured, it is possible to internalise the cost of such accumulation by means of a penalty on persistence. The idea here is to cause the design of chemicals to shift toward a level of persistence that balances both the chemical benefits of that trait as well as its accumulation-based costs. This implies the need for a value-based penalty on persistence, i.e. a quantitative penalty that recognises that persistence generates both benefits and costs for society. For this reason we suggest an instrument termed the 'accumulation tax'. The accumulation tax is a unit production tax that is equal to the product of: (1) the anticipated proportion of that unit of production that will accumulate ultimately within groundwater, and (2) the cost of an additional unit of chemical contamination in the groundwater resource.

What would determine the optimal level of the accumulation tax? The unit value of the water resource is equal to its 'opportunity cost', i.e. its value for its alternative use. In this context, the groundwater aquifer is being allocated between two uses: sink for chemical wastes or source for drinking water. The opportunity cost of groundwater contamination is the loss of drinking water purity and this may be estimated via the various methods introduced in Part II.

How does an accumulation tax address the problem of appropriately constructed regulation? In effect, this approach to regulation allows the chemical producer to choose between designing the chemical so that less of it will accumulate in the groundwater and designing the chemical for more accumulation with a penalty equal to the cost of each unit of groundwater use that design implicitly entails. Such an approach allows for regulation to achieve a balance between the costs of accumulation and the benefits that accumulative characteristics imply. It also gives the regulator a straightforward mechanism for shifting chemical products away from persistence: higher accumulation taxes will result in reduced overall accumulation in groundwater.

Taxing the implicit use of groundwater is an important step toward the rational regulation of this environmental resource; however, the problem of strategic response remains in regard to the other environmental media. The issue in this instance concerns the obvious avenues by which a firm might substitute other, untaxed resources if the tax on water resources is implemented. One such problem concerns the choice of the trait of affinity. If the use of groundwater is regulated, then this means that the relative incentives to exploit other unregulated media are enhanced. For example, the chemical might be designed for affinity with air rather than water, leading to accumulation in the atmosphere rather than in the groundwater. This suggests the need for an integrated accumulation tax, balancing the relative cost of contamination of various media.

Finally, it is important to note that much of this volume has been addressed to regulating chemical accumulation *vis à vis* the chemical industry, while there are admittedly many other agents who are able to determine the final cost of chemical accumulation (chemical users, water users, public water providers). This is not meant to imply that the problem is ultimately sourced with the chemical designer alone, but rather to pay some attention to a previously neglected facet of a complete chemical accumulation policy. Chemical characteristics such as persistence and affinity are endogenous to the production process, and thus subject to regulation. Properly constructed regulation must take this possible route of intervention into consideration.

Other routes to intervention are also available, e.g. chemical users. There is no question that the local users of chemicals are also important agents in determining the ultimate extent of chemical accumulation. An accumulation tax must be supplemented by other instructments that work through these other agents. The accumulation tax proposed here will necessarily apply to the average conditions existing across large swathes of territory, and perhaps, on account of economies of scale, the entire globe. Local communities with larger-than-average valuations of their groundwater supplies may wish to supplement this general producer-level accumulation tax with a supplementary local-level accumulation tax on users. Alternatively, they might not make use of their groundwater for any other purpose, and hence wish to subsidise the use of agricultural chemicals in their region. Clearly, the accumulation tax on producers might be supplemented by a completely different tax/subsidy schedule on users in all of the various regions in which it is applied.

The problem with a heterogeneous tax schedule within an area of ready mobility is, of course, the problem of enforcement. How can it be ensured that the users do not acquire all of their supplies from within the region with the most beneficial user tax schedule? Under such circumstances (as in the EU) where heterogeneity is important but mobility is mandatory, a supplementary layer of 'zoning' may be preferable to a supplementary tax schedule (Fauve and Lefevere, Chapter 10, this volume).

In essence, the overall objective of chemical regulation is to minimise the aggregate cost of: (1) agricultural losses due to weeds and weed control, (2) consumer risks and welfare losses due to water contamination, and (3) governmental costs due to administration and evasion. The first two objectives are balanced by means of an appropriately determined accumulation tax. The third objective may have to consider introducing alternative forms of instruments such as zoning or standards.

Conclusion

This volume is intended to demonstrate how science, economics and policy may all be integrated to address an important environmental problem. It is arguable that in this context the problem is insoluble in the absence of an integrated approach.

We demonstrate here that the EU's approach to groundwater regulation is based on a fundamentally flawed approach. It has attacked accumulation as an unmitigated bad thing rather than recognising it as the linked outcome of useful chemical production and application. It seeks to impose 'bans' rather than 'balance' and in the process achieves neither objective. Chemicals have continued to be manufactured and applied with the same in-built capacity for chemical accumulation. Groundwater contamination continues to escalate as a problem both in those areas where it is already present and increasingly where it has not been before.

Real and effective regulation of chemical accumulation requires an understanding of: (1) the characteristics of chemicals that contribute to accumulation, (2) an understanding of why they are demanded and by whom, (3) an even-handed objective that balances the benefits of these chemical characteristics with the costs of the accumulation that they imply, (4) an understanding of the values implicit in resource contamination, and

(5) an understanding of the instruments necessary to implement the desired balance between costs and benefits. This volume provides illustrations of how to think about and how to implement these various considerations. Most importantly it demonstrates the value of an integrated approach to policy-making in the area of environmental resources. It requires a collaborative effort between natural, social and policy scientists to bring the nature of the fundamental problem and the optimal policies to light.

References

Bergman, L. and Pugh, M. (eds.) (1994). *Environmental Toxicology, Economics and Institutions.* Kluwer Academic Publishers, Dordrecht.

Beach, D. and Carlson, G. (1993). A hedonic analysis of herbicides: do user safety and water quality matter? *American Journal of Economics,* **75**, 612–23.

Söderqvist, T. (1994). The costs of meeting a drinking water quality standard: the case of atrazine in Italy. In *Environmental Toxicology, Economics and Institutions,* ed. L. Bergman and M. Pugh, pp. 151–71. Kluwer Academic Publishers, Dordrecht.

Toman, M. and Palmer, E. (1997). How should an accumulative substance be banned? *Environmental and Resource Economics,* **9**, 83–102.

Part I

The characteristics of accumulative chemicals

2 Chemical characteristics: the case of herbicides in Italy

Carolina Sbriscia Fioretti, Giuseppe Zanin, Paolo Ferrario
and Marco Vighi

The risk of herbicides in groundwater

Chemical control in agricultural practice has a recent history, but it has increased greatly in the last few decades. At the end of the last century in Europe, copper sulphate was already being used to control broad-leaved weeds in grass crops, but it is only since the 1950s that selective herbicides have been introduced onto the market. The enormous success of these products was an incentive for research and development and this led to a number of new herbicides belonging to different chemical classes being marketed. This was followed by a big expansion of the use of chemical products for weed control worldwide. In developed countries, herbicides are used on 85% to 100% of all main crops. In the 1994 *Annual Index of Weed Abstracts*, 333 active ingredients are listed, and about 200 of these are widely distributed and used all over the world.

Although the environmental risk from pesticides was recognised relatively early on, the occurrence of pesticides in groundwater was not detected until much later and the major concern was directed towards DDT and other persistent organochlorine compounds. This was due mainly to lack of knowledge about the important features of chemicals movement through the soil, and gave rise to the general misconception that less persistent pesticides could not leach into the groundwater under normal conditions.

One of the first references to the discovery of pesticides other than chlorinated hydrocarbons in groundwater was by Richard *et al.* (1975). In the late 1970s, however, the number of detections increased rapidly, along with concern from public authorities about chemical contamination of groundwater.

Why herbicides?

According to a worldwide literature review by Funari *et al.* (1995), herbicides are the most frequently detected chemical pollutants in water, particularly in groundwater. This trait of herbicides can be easily explained by examining their uses and properties compared with other classes of widely used pesticides, e.g. insecticides and fungicides.

Total loads: Herbicides represent about 44% of the total pesticides market, compared with about 29% for insecticides, 21% for fungicides and 6% for other pesticides.

Mode of use and time of application: Herbicides are generally applied directly onto the soil or during the first stage of the cultural cycle, when the canopy of crop and weeds is limited. Other pesticides, on the other hand, with the exception of nematicides and insecticides used for soil disinfection, are usually sprayed onto plants. It can be hypothesised that very roughly the amount of sprayed pesticide reaching the soil surface could be between 20% and 50% of the total amount applied, and this significantly reduces the total load of a chemical to the soil system. Moreover, in normal agricultural practice, herbicides and nematicides are applied in a few treatments (normally one or two), at either pre- or post-emergence (see pp. 25–6), but in any case at the beginning of the growing season. This corresponds to the rainy season (autumn and spring) in temperate regions. Insecticide and fungicide treatments conversely are repeated for several months during the growing season. A significant amount of the load is applied in the dry season and it could be subject to several removal processes (degradation, volatilization, etc.) before being transported in surface or groundwaters; this is particularly true for insecticides. Wet conditions, however, could increase the need for fungicide treatment.

Physico-chemical properties: As explained by Vighi *et al.* (Chapter 4, this volume), the leaching capacity of an organic chemical depends on its physico-chemical properties, such as water solubility, soil absorption coefficient (K_{oc}), or persistence. Due to their particular mode of action, which is explained in more detail below, herbicides have a relatively high water solubility and low K_{oc} in comparison with insecticides and, to a lesser extent, fungicides. Therefore most herbicides show a relatively high leaching potential.

For all the reasons described above, herbicides are therefore typical contaminants of water and, in particular, of groundwater.

Herbicide mode of action and application patterns

If they are not contact herbicides, to have an effect the herbicides must first be absorbed by the plant and then transported to the target organs. Herbicide penetration can be through the leaves, roots, emerging shoots (hypocotyl, mesocotyl), the type of penetration determining the type of application.

Soil-applied herbicides

The herbicides that penetrate through the roots, called soil-applied herbicides or residual herbicides, are normally distributed in pre-emergence, i.e. before the crop and weeds have emerged, or in some cases incorporated into the soil at pre-sowing. This group of herbicides is characterised by:

(1) a suitable persistence in the soil, thus guaranteeing more or less protracted control of weeds that germinate throughout the growing season;

(2) apoplastic translocation. Apoplastically mobile herbicides follow the same pathway as water. They enter the xylem and are swept upwards with the transpiration movement of water and soil nutrients. They accumulate in the leaves where they normally carry out their toxic action by blocking photosynthesis. Chemically, they are generally weak bases, therefore their degradation in the soil is influenced by the soil pH: if the pH is low the chemicals in the protonated form, are adsorbed into the mineral colloids and cannot reach the roots. They are also more subject to chemical degradation via hydrolysis. The application rates must therefore be set according to the soil characteristics, in particular clay content and pH.

Some soil-applied herbicides may be absorbed by young shoots as they develop and grow upwards through the soil after seed germination. These herbicides are more or less persistent and differ from those which penetrate through the roots because:

(1) they are usually non-ionic molecules and as such are adsorbed mainly by organic matter with weak chemical bonding;
(2) they are not translocated far from the point of penetration;
(3) they have a prevalently graminicide action. Emerging shoots of grass weeds are very important sites of uptake of certain herbicides.

For both these groups of herbicides, efficacy is strongly affected by the soil moisture content: it is optimal in moist soil and reduced or nil in dry soil; satisfactory control requires an average of 10–15 mm of rainfall within 15–20 days after distribution.

Herbicide applied on vegetation

The herbicides that penetrate through the leaves are distributed after the crop and weeds have emerged (i.e. in post-emergence). According to the type of translocation, different groups can be distinguished.

(1) *Contact herbicides* are immobilised in the outer cell layers; their strong caustic action devitalises the surrounding cells, blocking translocation. For these herbicides the timing of the treatment is extremely important as their efficacy is greatly reduced if the weed is at an advanced stage.
(2) *Symplastic mobile herbicides* follow the same pathway as sugars formed in the leaves by photosynthesis. They move out of the leaf and go either up or down via the phloem, accumulating in underground stored organs or in those areas where sugar is used for growth. The direction of flow is governed by the location of leaves and growing points.
(3) *Ambimobile herbicides* enter in the phloem and can move from it to the xylem. Herbicides that move symplastically and migrate to the xylem can translocate up and down, whereas the reverse is generally unlikely (Zimdahl, 1993).

These herbicides, in particular those of the last two groups, rarely penetrate the strictly lipophilic barrier formed by the cuticle; they must therefore be distributed, with surfactants and wetting agents, in the form of salts or esters of their parent active ingredients. In fact, they are generally weak acids with intermediate lipophilicity expressed as the octanol/water partition coefficient (log K_{ow} = 0 to 3) (Bromilow *et al.*, 1990). The penetration of these herbicides is also influenced by weather conditions: periods of cold or extreme heat inhibit uptake as they increase the amount of epicuticolar waxes. Normally the persistence of this group of herbicides is lower than that of the residual products, although exceptions exist.

Weed control on maize crops

Economic impact of weeds in maize crops

The growing cycle of maize is during the spring and summer, when the weed flora is most aggressive; high temperatures and the ready availability of water and nutritive elements enhance the competitive action of the weeds. Zanin *et al.* (1992) demonstrated that in north-central Italy the average yield loss, in the absence of any kind of control, is between 34% and 38%, depending on the composition of the weed flora. The break-even point, expressed in percentage yield loss, varies from 5% to 12% according to the treatments used and is exceeded in 83% and 95%, respectively, of cases where *Sorghum halepense* is either present or absent. Weed control is therefore an essential and unavoidable operation for the farmer and can be carried out by different methods – mechanical, chemical, agronomic. The chemical method is presently the most widely used, often coupled with mechanical means.

Evolution of weed communities and weed control strategy in Italy

Chemical weed control in maize began in Italy in 1964, the year in which atrazine was introduced. Cropping methods also changed at that time, becoming ever more dependent on inputs from outside the farm, especially mechanical and chemical, and being based largely on monoculture. These new cropping methods produced a series of changes to the weed flora which have been well documented in floristic surveys. Table 2.1 shows the main events involved in maize cropping methods and the floristic changes that these produced.

Critical analysis of the recent past can help us to understand the changes taking place and to forecast the effects linked to the introduction of new herbicide families, new cropping methods or new legislative interventions.

Table 2.1. *Events that have affected the evolution of maize weed flora in the Po Valley*

Year	Event	Change to weed flora[a]
1964	Introduction of atrazine	
1968		Reduction in number of weed species
1970		Increase in frequency of panicoid grasses with high metabolisation rate of atrazine
1973	Introduction of alachlor	
1975		Spread of perennial broad-leaved species
1978	Introduction of metolachlor	
1979		Spread of atrazine-resistant biotypes
1983	Beginning of soybean cultivation	Spread of ruderal weeds
1986	First reduction on atrazine use	
1990	Ban of atrazine	
1992	Introduction of set-aside	
1993	Introduction of sulfonylureas	

Note:
[a] The dates of weed flora changes are only representative.

It is not fortuitous that modules on the history of weed science form part of Weed Science courses in American universities (Pearce and Appleby, 1992), even if it seems slightly odd to speak of 'history' in a discipline which is less than 50 years old. It is undoubtedly young, but the acceleration with which events have succeeded one another is formidable both in the evolution of the flora and in changes in herbicides and cropping methods.

The intense floristic evolution has forced farmers to re-think their control strategy every four to five years, adapting it to a target that has in the meantime been changed owing to the selection pressure exercised by different parts of the cropping technique, weed control in particular. Table 2.2 reports indices, calculated on the basis of available floristic surveys, that quantify the annual rate of evolution of the maize weed flora in northern Italy between 1962 and 1987 (Otto *et al.*, 1994).

The most rapid evolution was during the first period (1962–1971), when the annual rate of changed reached 4.1%; this signifies that each year 4.1 species out of 100 changed, with the disappearance of some species or the

Table 2.2. *Similarity and transformation coefficients, disappearance and transformation rate of maize weed flora between the floristic surveys made in the Po Valley during 1962–1987 (from Otto et al., 1994).*

	1962–1971	1971–1987
Interval (years) between subsequent surveys (n)	9	16
No. of new species (A)	23	15
No. of disappeared species (D)	27	23
No. of remaining species (C)	29	31
Similarity coefficient (R)[a]	51.80	57.40
Disappearance rate (T_r)[b]	5.40	2.70
Transformation coefficient (N)[c]	63.30	55.10
Transformation rate (T_n)[d]	4.10	2.80

Notes:
[a] $R = [C/(C + D)] \times 100.$
[b] $T_r = (100 - R)/n.$
[c] $N = [(A + D)/(A + D + C)] \times 100.$
[d] $T_n = (100 - N)/n.$

appearance of new ones. Farmers were thus forced to adapt the strategy of greater chemical control to maintain weed control effectiveness: this determined a greater use of herbicides, in terms of both quantity and number of treatments (Table 2.3). The evolution of the weed community has therefore had important consequences, both economic and environmental, and it is the key to understanding the choices of the farmers and the dynamics of the herbicide loads released into the environment.

Historically it is possible to identify three well-differentiated periods of weed control in maize:

(1) 1964–1986: period of atrazine use.
(2) 1987–1991: transition period between atrazine availability and the appearance of sulfonylureas.
(3) Post-1992: period with the availability of sulfonylureas effective against summer grass weeds at post-emergence.

The use of atrazine in Italy was widespread up to 1986: however, the floristic evolution suggested a continual adaptation of the method of use; in particular the application rates were gradually reduced by up to more than 50% and other more specific graminicides were combined with it in

Table 2.3. *Examples of typical weed treatments in different years showing changes in amounts (g/ha) and number of herbicides involved in maize weed control from the introduction of atrazine in the Po Valley, according to the floristic changes*

Years	Treatments	Rate (g/ha)	Total amount (g/ha)	No. of herbi- cides
1964	Atrazine	2000	2000	1
1973	Atrazine + alachlor	1000 + 2160	3160	2
1975	Atrazine + alachlor and 2,4-D	1000 + 2160 + 350	3510	3
1979	Atrazine + alachlor and bentazon	1000 + 2160 + 810	3970	3
1983	Atrazine + alachlor and dicamba	1000 + 2160 + 210	3370	3
1990	Metolachlor + pendimethalin + terbutylazine	1330 + 665 + 665	2660	3
1993	Rimsulfuron + dicamba	16 + 210	226	2

pre-emergence. The dynamics of the floristic changes then forced more intensive recourse to post-emergence herbicides to control specific groups of broad-leaved weeds, such as resistant biotypes (*Amaranthus* spp., *Chenopodium album*, *Solanum nigrum*, etc.), ruderal species (*Abutilon theophrasti, Xanthium* spp., etc.), or perennial species (*Calystegia sepium, Phitolacca americana*).

The local restrictions of the use and then the national ban of atrazine determined an adaptation in weed control strategy in Italy during 1987–1991 which can be summarised as follows:

(1) Recourse to a large number of herbicides, especially those used pre-emergence.
(2) Greater use of herbicides belonging to the amide and dinitroaniline families (alachlor, metolachlor, pendimethalin, etc.).
(3) Pre-emergence use of mixtures of two to three herbicides.

The introduction of sulfonylureas onto the Italian market in 1992 made weed control in maize possible with post-emergence treatments alone. Farmers, however, have been reluctant to shift weed control completely to post-emergence: in 1994 it was calculated that in Italy, on a total surface area of just over one million hectares of growing maize, around 230 000 hectares were treated with sulfonylureas at post-emergence alone. Given

that around 60 000 hectares are peaty soils where treatment at post-emergence is compulsory, the surface area gained by this intervention strategy is
limited to the potentially treatable hectares. These should include at least
all areas infested by *Sorghum halepense* and be extended to soils less
infested with summer grass weeds. Refinements in the method of use (in
particular formulations of specific mixtures for the various infestations,
definition of the optimal treatment time, better understanding of the
potential of the sulfonylureas) will favour the expansion of this strategy in
the future.

Recently, a possible future strategy based on a compromise between the
old strategy of pre-emergence treatment and the new post-emergence
treatment has been put forward. This is based on a mixture of a residual
herbicide, such as terbutylazine, and a sulfonylurea at early post-emergence (maize at two to three leaves). In soils not heavily infested with perennial grass weeds, this solution could be optimal.

Time and space distribution of atrazine bans in Italy

The use of atrazine in the years 1987 to 1990 was severely affected by the
local bans issued by the Regional Administrations of the Po Valley.
Moreover, in these years the maximum application rate allowed was
reduced at national level. It is easy to understand the confusion and the
difficulties encountered by the farmers, who, in some cases, had to 'invent'
new combinations of herbicides.

The Directive 80/778/EEC, relating to the quality of water intended for
human consumption, establishes maximum admissible levels of chemical
and micro-organisms in drinking water. It also fixes methods and frequency of analysis for control (Faure, 1994). The Directive has been implemented in Italy with the DPCM (Decreto del Presidente del Consiglio dei
Ministri) of 8 February 1985. In compliance with the law, control analyses of
groundwater were carried out. As a consequence in many areas of the Po
Valley, where the majority of the Italian maize crop is located, the levels of
atrazine and other herbicides frequently exceeded the EU standard for pesticides of 0.1 $\mu g/l$ (Vighi and Zanin, 1994). A number of regional regulations
were issued in Piedmont, Lombardy, Emilia-Romagna and Veneto from
1987 to 1990 either for derogation to a temporary drinking-water standard
(toxicologically safe but less stringent than the EU limit) or for limiting the
use of several herbicides in the municipalities where high values of these

Table 2.4. Maize crop areas (ha) and areas (ha) where atrazine was banned in the Po Valley between the years 1987 and 1990

	1987		1988		1989		1990	
	Maize crop area	Banned area	Maize crop area	Banned area	Maize crop area	Banned area	Maize crop area	Banned area
Veneto	181 200	26 408	209 208	28 387	261 312	80 484	260 736	81 425
Piedmont	128 350	0	156 906	20 188	195 984	Total ban[a]	195 552	Total ban[a]
Lombardy	143 450	40 487	174 340	40 487	217 760	90 727	217 280	90 727
Emilia-Romagna	37 750	0	43 585	0	65 238	0	76 048	0
Total	490 750	66 895	584 039	89 062	740 384	367 195	749 616	367 704

Note:
[a] In the case of total ban, the banned area corresponds to the maize crop area.

compounds had been found. In particular, the ban involved the herbicide atrazine, which at that time, was used largely on the maize crop. Among the four regions, only three (Lombardy, Veneto and Piedmont) issued regulations that greatly limited the use of atrazine before the national ban implemented in 1991.

In Piedmont, the OPGR (Ordinanza del Presidente della Giunta Regionale) of 17 April 1987 and 13 April 1988 forbad the use of atrazine in the municipalities where groundwater analysis showed levels above 1 μg/l. With the 'Delibera del Consiglio Regionale del Piemonte' of 1 March 1989, the sale and use of each kind of herbicide containing atrazine was forbidden. On 17 April 1987 an OPGR in Veneto prohibited the use of atrazine in those municipalities in which the level monitored in groundwater exceeded 1 μg/l. The restriction was then repeated up to May 1990. In Lombardy similar restrictions were issued on 27 April 1987, and 4 April 1989. In Emilia-Romagna, although a restriction was issued in April 1988, levels of atrazine exceeding 1 μg/l were not detected in the regional territory so the ban did not become executive.

The whole situation relating to local bans in the years 1987 and 1990 is shown in Table 2.4 and Figures 2.1, 2.2 and 2.3.

Selection criteria for alternative chemicals

The banning of atrazine had agronomic and economic consequences in the agriculture of northern Italy. In the four major regions of the Po Valley, the maize crop represented about 12.5% of the total UAA (utilised agricultural area) and weed control of this crop was mainly with atrazine. There was therefore an urgent need to find substitute chemicals. The selection criteria adopted by the farmers for the choice of herbicides were examined in a recent survey (Infomark, 1993). In Table 2.5 the 13 major factors influencing the selection of herbicide by farmers are listed in order of importance.

Among these factors, seven refer to agronomic aspects (absence of residues, selectivity, action spectrum, persistence, speed of action, specific action on grass weeds or specific action on dicotyledons), the others relate to economic reasons, safety, and direct (personal experience) or indirect (retailer advice, company assistance) information. The economic aspect plays an important role, demonstrated not only by the high score of the specific item (costs) but also by the indirect economic fall-out of other

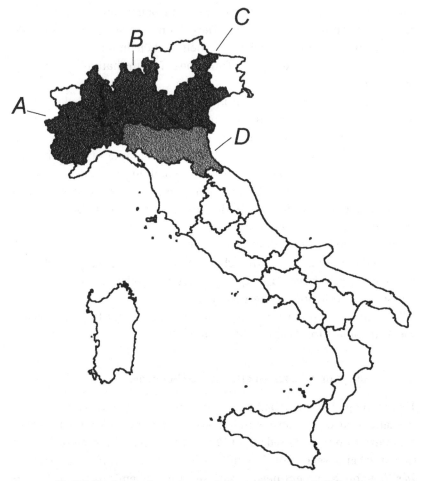

Fig. 2.1. The four major maize-producing regions of the Po Valley: A, Piedmont; B, Lombardy; C, Veneto; D, Emilia-Romagna.

items. For example, the Italian trade union regulations impose a half day of rest for each half day of work with highly toxic compounds (toxicological class I) and this produces a significant increase in cost for large farms. Moreover, ease of use affects the overall cost of a pesticide treatment (time needed to prepare formulation, time and structure needed for spreading, etc.). In order to better understand the real meaning of some of the factors listed in Table 2.5 and to explore the possibility of evaluation or quantification, we need to have some agronomic, environmental and toxicological explanations.

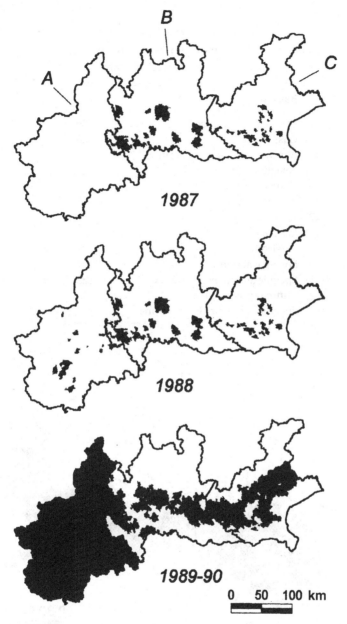

Fig. 2.2. Areas involved in the atrazine ban in the period 1987–1990. For key, see Figure 2.1.

Table 2.5. *Reasons for choosing a product (Infomark, 1993). % indicates the proportion of positive answers in a sample of 319 farmers*

Reason	%
(1) Personal experience	69
(2) Costs	55
(3) Absence of toxic residues in soil	54
(4) Selectivity	42
(5) Toxicological class	41
(6) Action spectrum	38
(7) Retailer advice	35
(8) Ease of use	34
(9) Persistence	29
(10) Speed of action	24
(11) Specific action on grass weed	22
(12) Company assistance	15
(13) Specific action on dicotyledons	7

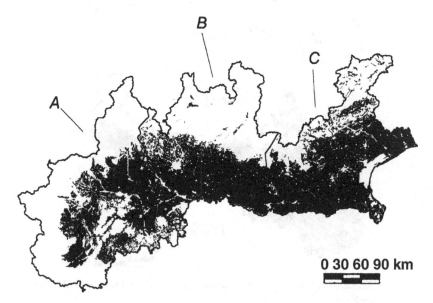

Fig. 2.3. Utilised agricultural area (UAA) in the three regions involved in the atrazine ban. This figure was compiled using ARC/INFO software. For key, see Figure 2.1.

Agronomic persistence

The problem of persistence, from an environmental point of view, is related only to the degradation and transformation processes of a compound. It does not take into account transport processes from one environmental compartment to another (e.g. from soil to water) as opposed to a genuine disappearance from the overall environment (see Vighi *et al.*, Chapter 4, this volume). From an agronomic view point, persistence is the resistance to either the degradation or the transfer processes. Agronomic persistence is not a property defined by the nature of the herbicide alone; it varies according to climate, soil type, method of use (application rate, distribution time, etc.) and tillage system. The farmer is usually interested merely in the agronomic persistence, i.e. the time during which there is a phytotoxic effect on the weeds and/or crop. This depends on the herbicide concentration in the soil solution, which is regulated by adsorption. Clearly, agronomic persistence varies from situation to situation. And index used to evaluate persistence in the soil is the half-life (DT_{50}, the number of days necessary for the herbicide concentration to halve). This, if calculated under field conditions, includes both degradation phenomena and removal (run-off, volatilisation, leaching, etc.). The half-life in the soil affects rows (3) and (9) of Table 2.5, but the implications are different. Absence of toxic residues in the soil (row (3)) avoids carry over effects to the following crop in a rotation plan, but row (9) (persistence) is related to prolonged efficacy of the compound.

Selectivity and action spectrum

Selectivity is the ability of a herbicide, intrinsic or induced, to impact upon non-target species in a differential manner: the non-target species (i.e. the crop) should resist the toxic action of the chemical while the target (weed) species succumb. Herbicides can be divided into non-selective and selective, according to their action spectrum: the former control almost all weeds while the latter have no effect on certain species, which is of varying importance. Selective herbicides are most often used in agriculture as the crop must be safeguarded, but non-selective herbicides are used in some situations such as seedbed cleaning or weed control in orchards. No herbicide belongs rigidly to either group because selectivity is primarily a function of application rate.

Three different types of selectivity exist (Sattin *et al.*, 1995):

(1) Selectivity by lack of, or minimum interaction between, plant and herbicide; it happens through separation in time and space, so that there is negligible or no contact between the two. This could result from the distribution method, plant morphological characteristics or different positioning of the herbicide and root system through the soil profile. The last of these, also called positional selectivity, is strongly influenced by the amount of rainfall. Positional selectivity is complete when a poorly soluble herbicide with a high K_{oc} remains in the surface layers of soil, while the root system of the crop is found deeper down. Many good pre-emergence herbicides exploit this type of selectivity, which greatly depends on the pedo-climatic factors that regulate both the movement of the herbicide through the soil profile and the crop growth rate. Rainfall in the first two to three weeks after sowing is important because, after the crop's roots have penetrated deeper down, the possibility of damage is reduced.

(2) Physiological selectivity is when herbicide and plant interact completely. It happens through different mechanisms that either prevent the herbicide from reaching the target site (translocation selectivity) or detoxify it before it reaches the target site (detoxification selectivity). Selectivity can also depend on the insensitivity of the target site itself. For example, diphenamid acts on the roots by inhibiting growth: in sensitive plants there is slow translocation towards the photosynthetic tissues, the only site where detoxification by combination with glucose takes place.

(3) Selectivity can be induced by disrupting the herbicide metabolism or by manipulation of crop tolerance. It is obtained with the use of 'safeners', i.e. substances that selectively protect crop plants against herbicide injury by altering the normal plant metabolism and improve herbicide detoxification. Genetic engineering is also used to introduce, into the crop, specific genes coding for an altered target site or for an enzyme involved in herbicide detoxification. This produces resistant crops, also known as transgenic plants.

Generally, some selective herbicides have a prevalent action against grass weeds and others are more effective in the control of dicotyledons.

Herbicides effective against perennial weeds must then be distinguished from those active solely against annual species. The presence of species belonging to both these groups often forces the farmer to choose herbicide mixtures, sometimes even complex ones. The action spectrum of a herbicide can be measured quantitatively by means of the so-called kill rate, defined as the percentage of weed control obtained through an application. The kill rate of a herbicide alters with time, as the botanical composition of the weed community changes. These properties are important criteria affecting the farmer's choice of chemical (see rows (4), (6), (11) and (13) in Table 2.5). The farmer must combine knowledge of the weed flora present and information on the herbicide action spectrum.

Reliability

Reliability is the ability of a herbicide to maintain its biological action, given the variation of pedo-climatic conditions and growth stage of the weeds. It is a characteristic of enormous agronomic importance. The reliability of a herbicide in different agronomic and environmental conditions makes the compound more easy to use (row (8) of Table 2.5). In general, farmers derive their information on reliability from personal experience.

Toxicological class

According to the Italian law on pesticides (DPR 223/88) toxicity is classified on the basis of the lethal dose 50 (LD_{50}, see p. xiii) via either oral ingestion or cutaneous exposure. LD_{50} is measured as milligrams of active ingredients for each kilogram of body weight. The greater the value of the LD_{50} the lower is the toxicity of the corresponding active ingredient. This toxicological classification takes into account only acute toxicity and is intended primarily for the protection of farmers from the risk of contamination due to direct exposure to pesticides during manipulation and use. In Italy, pesticides are subdivided into groups based on the two toxicological parameters (oral and cutaneous) (see Table 2.6).

Application of the selection criteria to atrazine substitutes

A system of classification of herbicides based on all the different reasons given by farmers for choosing a product is practically impossible. Most of

Table 2.6. *Toxicological classes of pesticides, according to Italian law*

	Solid LD$_{50}$ (mg/kg)		Liquid LD$_{50}$ (mg/kg)		Gaseous LD$_{50}$ (mg/kg)
	Oral	Cutaneous	Oral	Cutaneous	Inhalation
From I Class					
Very toxic	≤ 5	≤ 10	≤ 25	≤ 50	≤ 0, 5
Toxic	5–50	10–100	25–200	50–400	5–50
From II Class					
Harmful	50–500	100–1000	200–2000	400–4000	2–20
From III Class Irritant or not classified	> 500	> 1000	> 2000	> 4000	> 20

the factors shown in Table 2.5 cannot be precisely quantified. In many cases they are only qualitative statements or are related to personal, subjective feelings of the farmers. In other cases, they indicate properties of the herbicides which can be objectively evaluated but which are highly variable in time and space in their effect on environmental characteristics (climate, weed community, etc.). Nevertheless an attempt to quantify some of these properties has been made.

In order to produce the information needed for a hedonic analysis of the demand for pesticide characteristics (Söderqvist, Chapter 3, this volume), five properties of the herbicides, for which a continuous quantification was possible, have been selected: price, persistence, action spectrum, reliability and toxicity. These properties cover approximately 10 out of 13 of the factors indicated in Table 2.5. Quantitative information for the five properties has been provided for the alternatives to atrazine, in order to provide input data for economic modelling (Söderqvist, Chapter 3, this volume).

Price

Prices of the different active ingredients and their yearly variations have been collected for the period 1987–1994. Data supplied for the economic evaluation are not reported here.

Table 2.7. *Half-life (DT_{50}) of the weed control treatments on the maize crop in north-east Italy*

Treatments	kg formulated product/ha	DT_{50} (days)
Pre-emergence		
(1) Atrazine	3	80
(2) Atrazine + alachlor	2 + 5	38
(3) Atrazine + metolachlor	2 + 2	47
(4) Atrazine + EPTC or butylate	2 + 7.5	23
(5) Pendimethalin + atrazine	3.5 + 2	62
(6) Terbutylazine + metolachlor	2 + 3	43
(7) Terbutylazine + alachlor	2 + 4.5	34
(8) Alachlor + linuron	4 + 1	28
(9) Metolachlor + linuron	2 + 1	38
(10) Pendimethalin + alachlor	2 + 4.5	28
(11) Pendimethalin + metolachlor	2 + 2.5	37
(12) Alachlor	5	20
(13) Metolachlor	3	30
(14) Linuron + pendimethalin	0.7 + 2	54
Post-emergence		
(15) Rimsulfuron	0.055	12
(16) Primisulfuron	0.025	40
(17) Rimsulfuron + dicamba	0.055 + 0.8	14
(18) Primisulfuron + dicamba	0.025 + 1	16
(19) Rimsulfuron + terbutylazine	0.055 + 2.2	69

Persistence

The agronomic half-life is extremely variable, mainly as a result of environmental factors (temperature, moisture, rain, presence of adapted bacteria, etc.). A survey of the literature and personal experience have allowed the authors to propose 'reasonable' values, assumed to be an acceptable approximation of the behaviour under 'normal' conditions in northern Italy during different herbicide treatments (Table 2.7). For herbicide mixtures, a weighted mean has been calculated, taking into account the half-life (DT_{50}) of single components and the amount of each active ingredient. The sulphonylureas have little or no influence on the DT_{50} values of post-emergence mixture treatments because of the low percentage present in

Table 2.8. *Kill rate (% of control) of weed control treatments on the maize crop, calculated for two different periods*

	Kill rate	
Treatments	1987–1988	1989–1994
Pre-emergence		
(1) Atrazine	62.3	57.5[a]
(2) Atrazine + alachlor	79.7	74.9[a]
(3) Atrazine + metolachlor	79.7	74.9[a]
(4) Atrazine + EPTC or butylate	83.0	
(5) Pendimethalin + atrazine	75.9	
(6) Terbutylazine + metolachlor	80.1	74.4
(7) Terbutylazine + alachlor		74.4
(8) Alachlor + linuron	79.0	73.8
(9) Metolachlor + linuron	79.0	73.8
(10) Pendimethalin + alachlor	77.1	73.1
(11) Pendimethalin + metolachlor	77.1	73.1
(12) Alachlor	71.7	67.3
(13) Metolachlor	71.7	67.3
(14) Linuron + pendimethalin		73.6
Post-emergence		
(15) Rimsulfuron		84.6
(16) Primisulfuron		84.8
(17) Rimsulfuron + dicamba		92.0
(18) Primisulfuron + dicamba		87.3
(19) Rimsulfuron + terbutylazine		92.0

Note:
[a] Years 1989–1990.

the mixture. It must be stressed, however, that the dose per hectare used in post-emergence is about half that used for pre-emergence treatments.

Action spectrum

The average kill rates of certain herbicides or mixtures of herbicides (Table 2.8) have been calculated for two different periods (1987–1988, weed community typical of the period before the ban on atrazine; 1989–1994, weed community typical of the transition period and of period with sulphonylureas available). Within each period, the frequency of each species has been multiplied by the respective kill rate. The sum obtained was then divided by the number of species present.

The different incidences of the species frequency during both periods and the different kill rates of the herbicides produced differences at various treatment levels. The shift in weed flora composition reduced the global kill rate of different mixtures of herbicides used before the atrazine ban. The introduction of sulphonylureas, alone or as mixtures, increased the weed control after the change in the weed community.

Reliability

In residual herbicides, reliability is increased the more the chemical is released from the soil by moisture: that is to say, the more soluble and persistent herbicide is, with low soil retention, the more reliable it is. The GUS index (groundwater ubiquity score; Gustafson, 1989) is based on the organic carbon sorption coefficient (K_{oc}) and disappearance time in the soil (DT_{50}), and was originally developed to describe the potential for groundwater contamination (see Vighi *et al.*, Chapter 4, this volume). It could, however, also represent a screening approach to evaluate the reliability of soil-applied herbicides: a high GUS signifies high mobility and so better reliability, a low soil moisture content being enough to activate the herbicide. Values of GUS in excess of 2.8 indicate good reliability, whereas values below 1.8 indicate poor reliability. Table 2.9 reports GUS values of pre-emergence treatments, using an arithmetic mean for the mixtures.

Reliability is more difficult to define for post-emergence herbicides. It can, however, be concluded that a post-emergence herbicide is all the more reliable the wider its flexibility of use and the more complete the action spectrum. A herbicide is defined as 'flexible' if it also maintains its efficacy on developed weeds, i.e. if it can be used effectively over a long period. Flexibility is wide in the symplastic mobile herbicides and in the ambimobile ones, whereas in contact herbicides efficacy is strongly linked to the growth stage. These herbicides are effective only on weaker weeds; with more developed, stronger weeds, effectiveness declines drastically. The action spectrum is well differentiated in the different groups: some herbicides are active only on annual dicotyledons, others on perennial dicotyledons, and yet others control a good number of both dicotyledons and grass weeds. To calculate reliability for post-emergence treatments, scores have

Table 2.9. *GUS index of pre-emergence treatments (<1.8, low reliable; 1.8–2.8, fairly reliable; >2.8, very reliable)*

Pre-emergence treatments	GUS index
(1) Atrazine	2.89
(2) Atrazine + alachlor	2.35
(3) Atrazine + metolachlor	2.63
(4) Atrazine + EPTC or butylate	2.40
(5) Pendimethalin + atrazine	0.785
(6) Terbutylazine + metolachlor	2.74
(7) Terbutylazine + alachlor	2.46
(8) Alachlor + linuron	2.75
(9) Metolachlor + linuron	3.03
(10) Pendimethalin + alachlor	0.240
(11) Pendimethalin + metolachlor	0.525
(12) Alachlor	1.80
(13) Metolachlor	2.37
(14) Linuron + pendimethalin	1.19

been assigned to flexibility of use and to the effectiveness against different groups of weeds (annual dicotyledons, perennial dicotyledons, annual grass weeds and perennial grass weeds). The scoring for the calculation of the reliability index for post-emergence and the mean values for the different treatments are reported in Tables 2.10 and 2.11.

Toxicity

In order to have continuous ranking, the toxicity of products used for weed control on the maize crop has been expressed as LD_{50} (mg/kg) values for oral ingestion (Table 2.12). For the herbicides mixed by the farmer, the highest toxicity value has been reported (lowest LD_{50}). For the products sold as mixtures, the values have been calculated taking into account the different percentages of active ingredient per product unit.

Post-emergence treatments generally show lower toxicity. This is better illustrated in Table 2.13, where the toxicity of pre- and post-emergence treatments is compared. This is in line with the general trend of development of new active ingredients in pesticide research, aimed at producing

Table 2.10. *Scoring for the calculation of the reliability index (RI) of post-emergence treatments*

	Flexibility (score)	Action spectrum (score)
Very high	1	5
High	2	4
Medium	3	3
Low	4	2
Very low	5	1

Table 2.11. *Scoring and reliability index (RI) of post-emergence treatments (<3, low reliable; 3–4, fairly reliable; >4, very reliable)*

Post-emergence treatments[a] (RI = A + B/2)	(A) Flexibility (score)	(B) Action spectrum (score)	Reliability index
(15) Rimsulfuron (1)	4	3	3.5
(16) Primisulfuron (1)	4	2	3.0
(17) Rimsulfuron (1) + dicamba (2)	5	4	4.5
(18) Primisulfuron (1) + dicamba (2)	5	3	4.0
(19) Rimsulfuron (1) + terbutylazine (3)	4	4	4.0

Note:
[a] (1), Ambimobile herbicide; (2), Symplastic mobile herbicide; (3), Apoplastic mobile herbicide.

compounds with increasing specific efficacy and decreasing harmfulness to humans and the environment.

Conclusions

From the foregoing, it appears that the choice of herbicide is a complex problem that involves agronomic and environmental aspects. The 'correct set of characteristics' of a hypothetical ideal herbicide should include farmers' requirements (listed in Table 2.5) and optimal properties for protection of

Table 2.12. *LD_{50} (mg/kg) of weed control treatments on the maize crop*

Treatments		Toxicity LD_{50}
Pre-emergence		
(1) Atrazine		5500
(2) Atrazine + alachlor	A	3320
(3) Atrazine + metolachlor	A	5930
(4) Atrazine + EPTC or butylate	B	3660
(5) Pendimethalin + atrazine	A	2930
(6) Terbutylazine + metolachlor	A	5220
(7) Terbutylazine + alachlor	B	2570
(8) Alachlor + linuron	B	2570
(9) Metolachlor + linuron	A	6930
(10) Pendimethalin + alachlor	B	2570
(11) Pendimethalin + metolachlor	A	3750
(12) Alachlor		2570
(13) Metolachlor		4060
(14) Linuron + pendimethalin	A	5820
Post-emergence		
(15) Rimsulfuron		20 000
(16) Primisulfuron		6730
(17) Rimsulfuron + dicamba	B	8050
(18) Primisulfuron + dicamba	B	6730
(19) Rimsulfuron + terbutylazine	B	4850

Note:
A, sold as a mixture; B, mixed by farmer.

natural resources, in particular groundwater (low mobility, low leachability, low persistence, etc.). The problem is highly controversial because the 'good' or 'bad' properties of a herbicide are not unequivocally defined. A relatively high persistence can be considered a positive or negative factor by the farmer(see p. 37), but is, in any event, a negative factor for the environment. Selectivity is a good property for the farmer and the environment, but a high action spectrum, with a high kill rate, is a required agronomic property. Reliability is an important agronomic property, but, at least for pre-emergence herbicides, the more a compound is reliable the more it is leachable. Herbicide properties should therefore be carefully evaluated in

Table 12.13. *Minimum, maximum and mean values of LD_{50} (mk/kg) in pre-and post-emergence treatments*

	Minimum value LD_{50}	Mean value LD_{50}	Maximum value LD_{50}
Pre-emergence	2570	4100	6930
Post-emergence	4850	9270	20000

order to attain the best compromise between the requirements of agricultural production and those of environmental protection.

In general, farmers have the information and the skill needed to select the best treatments according to agronomic conditions (crop, climate, weed composition, etc.); nevertheless, technological progress (introduction of new chemicals, the need for mixtures of products, the addition of 'safeners' or additives) makes the farmer's task increasingly more complex. On the other hand, it is completely unrealistic to imagine that farmers could also have the skill necessary to select 'environmentally friendly' herbicides. To overcome these difficulties, computerised tools have recently been developed that can aid the farmer in herbicide choice. Most of these systems (Wilkinson *et al.*, 1991; Black and Dyson, 1993; Forcella *et al.*, 1993; Castro-Tendero and Garcia-Torres, 1996) identify, for the different weed communities, the best solutions in relation to the economic balance of the farm. The GESTINF program, developed by Berti and Zanin (1996), evaluates for each treatment not only the cost advantage but also an index of danger for the environment, called the groundwater danger index, GWDI (Berti *et al.*, 1995*a*). This program has been validated by trials on wheat and soybean. Farmers have therefore two criteria for choosing: cost and environmental friendliness. This situation is of great interest because often the most costly effective solutions for the farmer produce very different risks for groundwater (Berti *et al.*, 1995*b*).

The need for computerised support systems is increasing as a result of the growing complexity of agronomic and technological variables. In a survey carried out in the USA (Stoller *et al.*, 1993), the availability of these computerised systems was considered to be a priority for contractors, who are the key personnel in the management of weed control at ground level.

References

Berti, A. and Zanin, G. (1997). GESTINF: a decision model for postemergence weed management in soybean (*Glycine max* (L.) Merr.). *Crop Protection*, **16**, 109–16.

Berti, A., Zanin, G., Otto, G., Trevisan, M. and Capri, E. (1995*a*). Evaluation of the relationship cost–risk of groundwater contamination in weed control of soybean. *European Journal of Agronomy*, **4**, 491–8.

Berti, A., Sartorato, I., Zanin, G. and Sattin, M. (1995*b*). Valutazione economica e ambientale di diverse strategie di controllo delle infestanti: un approccio modellistico. *Rivista di Agronomia* **29**, 3 suppl., 331–8.

Black, I. D. and Dyson, C. D. (1993). An economic threshold model for spraying herbicides in cereals. *Weed Research*, **33**, 279–90.

Bromilow, R. H., Chamberlain, K. and Evans, A. (1990). Physico-chemical aspects of phloem translocation of herbicides. *Weed Science*, **38**, 305–14.

Castro-Tendero, A. J. and Garcia-Torres, R. (1996). SEMAGI – An expert system for weed control decision making in sunflowers. *Crop Protection*, **14**, 543–8.

Faure, M. G. (1994). The EC Directive on drinking water: institutional aspects. In *Environmental Toxicology, Economics and Institutions – The Atrazine Case Study*, ed. L. Bergman and M. Pugh, pp. 39–88. Kluwer Academic Publishers, Dordrecht.

Forcella, F., Buhler, D. D., Swinton, S. M. and King, R. P. (1993). Field evaluation of a bioeconomic weed management model for the corn belt, USA. In *8th EWRS Symposium on Quantitative Approaches in Weed and Herbicide Research and their Practical Application*, pp. 775–60. European Weed Research Society, Braunschweig.

Funari, E., Donati, L., Sandroni, D. and Vighi, M. (1995). Pesticide levels in groundwater: value and limitation of monitoring. In *Pesticide Risk in Groundwater*, ed. M. Vighi and E. Funari, pp. 3–44. CRC, Lewis Publishers Inc., Chelsea.

Gustafson, D. I. (1989). Ground water ubiquity score: a simple method for assessing pesticide leachability. *Environmental Toxicology and Chemistry*, **8**, 339–57.

Infomark (1993). Market Survey, courtesy of Dupont de Nemours and Co. Inc.

Otto, S., Zanin, G., Rapparini, R. and Mundula, S. (1994). La flora infestante estiva del mais in Pianura Padana. *L'Informatore Agrario*, **42**, 71–6.

Pearce, R. B. and Appleby, A. P. (1992). Survey of beginning weed science courses. *Weed Technology*, **6**, 1031–6.

Richard, J. J., Junk, G. A., Avery, M. J., Nehring, N. L., Fritz, J. S. and Svec, H. J. (1975). Analysis of various Iowa waters for selected pesticides: atrazine, DDE and diendrin – 1974. *Pesticide Monitoring Journal*, **9**, 117–23.

Sattin, M., Berti, A. and Zanin, G. (1995). Agronomic aspects of herbicide use. In *Pesticide Risk in Groundwater*, ed. M. Vighi and E. Funari, pp. 45–72. CRC, Lewis Publishers Inc. Chelsea.

Stoller, E. W., Loyd, M. W. and Alm, D. M. (1993). Survey results on environmental issues and weed science research priorities within the corn belt. *Weed Technology*, **7**, 763–70.

Vighi, M. and Zanin, G. (1994). Agronomic and ecotoxicological aspects of herbicide contamination of groundwater in Italy. In ed. L. Bergman and M. Pugh *Environmental Toxicology, Economics and Institutions – The Atrazine Case Study*, pp. 111–25. Kluwer Academic Publishers, Dordrecht.

Wilkerson, G. C., Modena, S. A. and Colbe, H. D. (1991). HERB: decision model for post-emergence weed control in soybean. *Agronomy Journal*, **83**, 413–17.

Zanin, G., Berti, A. and Giannini, M. (1992). Economics of herbicide use on arable crops in North-central Italy. *Crop Protection*, **11**, 174–80.

Zimdahl R. L. (1993). *Fundamentals of Weed Science*. Academic Press Inc., San Diego, CA.

3   Valuing chemical characteristics: a hedonic approach[1]

Tore Söderqvist

Introduction

In a seminal paper, Lancaster (1966) suggested a new approach to consumer theory. His suggestion was based on the idea that the satisfaction derived from the consumption of a good is not due to the good per se, but rather from the *characteristics* of the good. The hedonic analysis was developed further by, for example, Rosen (1974) and has been employed extensively for various purposes. For example, a hedonic method for estimating people's willingness to pay for a change in the provision of environmental quality characteristics has been used (with varying degrees of success) in environmental economics since the end of the 1960s.[2] The importance of product characteristics for consumers' enjoyment was, however, acknowledged within economic research at an early stage. In an example concerned with automobiles, Court (1939) constructed price indexes that accounted for quality change by following a procedure which he called the hedonic pricing method. A crucial part of the method was the identification of measureable characteristics that relate to automobile quality.[3]

It is interesting to note that a focus on characteristics may offer a procedure to predict the use of new products. Roughly speaking, it may be possible to describe a product which is not yet introduced into the market as a new combination of characteristics that describe already existing products. This possibility is not fruitful for some groups of products. For example, one important reason for buying a mug is probably its look, maybe the presence of a text or a picture on the mug. One or several 'look characteristics' that are likely to be related to people's purchase decisions are in such a case not easy to define or to measure. This chapter, however, deals with herbicides, a group of products for which objectively measurable characteristics are likely to be of a considerable importance for decisions about production and consumption. One example of such a characteristic is the kill rate of a herbicide. However, as is discussed in detail below, other characteristics based on the chemical properties of herbicides also are likely to be of importance for users' choice of herbicides. The importance of non-chemical properties is also examined.

As long as new herbicides (or new versions of existing herbicides) can be described by the same set of characteristics, a focus on this set could be relevant for predicting the use of the new herbicides. However, a technical revolution would probably make the existing set of characteristics obsolete. For example, the use of genetic engineering may result in crop varieties that are resistent to one or more herbicides (cf. Zanin *et al.*, 1995). Such a change in weed control technology would probably call for characteristics different from those which described the traditional herbicides. However, the herbicides that are considered in the empirical work described in this chapter are the main treatment alternatives in maize cultivation in northern Italy used over a rather short period of time (1987–1994). This period contained drastic legislative restrictions on the use of one important herbicide in Italian maize cultivation (atrazine) (see Bergman and Pugh, 1994, on the extent, causes and effects of these restrictions). However, at the same time there were no widespread revolutions in weed control technology such as that referred to above.

The Italian legislation on atrazine resulted from the appearance of this substance in groundwater, which is the dominant source of drinking water in northern Italy. More than 80% of the inhabitants in this region are likely to receive their tapwater from groundwater. Interestingly, some of the herbicide characteristics that are likely to influence the farmers' choice of herbicides for reasons related to maize production are also of importance to the risk of damage to human health and the environment from herbicide use. This fact suggests that a hedonic analysis may indicate whether user safety and water quality matter to farmers when they choose a herbicide. In order to study this particular issue, Beach and Carlson (1993) made a hedonic analysis of herbicides based on data from the USA. Some close parallels exist between their study and the present one. For example, their study also concerned maize (and soybean) cultivation and the definitions of characteristics are partly identical. This means that some interesting comparisons of empirical results can be made below.

In the following section, a model of herbicide choice is outlined. Data presented in the third section are used in the empirical work, i.e. an estimation of a hedonic price function, presented in the fourth section. The last section closes the chapter with a discussion and some conclusions.

A model

There are a number of different herbicide treatment alternatives available for weed control in maize cultivation. Such treatments, denoted here by h_i $(i = 1, \ldots, j, \ldots n)$, may involve one single herbicide or sometimes a combination of herbicides that is available commercially pre-mixed or is mixed by farmers before application. A farmer's choice of one of the n available treatments depends on agronomic, economic and climatic conditions, soil composition, weed community, etc. The situation where a farmer chooses the jth treatment is denoted by $\{h_j = 1, h_i = 0, \forall i \neq j\}$.

Assume that all treatments can be described perfectly by a vector of characteristics $z = [z_1, \ldots, z_k, \ldots, z_r]$. The r characteristics take different values for different treatments, which implies that each treatment corresponds to a unique combination of characteristic values. For example, the jth treatment, i.e. h_j, is described by the vector $z_j = [z_{1j}, \ldots, z_{kj} \ldots, z_{rj}]$. The empirical nature of these characteristics is discussed thoroughly in Chapter 2. For the moment, suffice it to note that some conditions of importance for the possibility of applying a hedonic model are satisfied (cf. Rosen, 1974, Beach and Carlson, 1993). First, in the section on data, market-oriented characteristics are defined, i.e. they will be perceptible to farmers, and supposedly significant for farmers' purchase decisions. Moreover, there are at present at least 20 distinct treatments available in the Italian maize herbicide market at various prices. This means that herbicide producers supply quite a wide variety of bundles of characteristics. Secondly, the possibility for a farmer to mix different herbicides (at the cost of some extra work) implies that the choice among packages of characteristics is to a large extent continuous.

Let us assume that farmers follow profit-maximising behaviour. Owing to, for example, unobserved variations in farmers' production processes, profits are defined as:

$$\pi = \nu + \epsilon \tag{3.1}$$

where π is profits per hectare, ν is the systematic profit component and ϵ is a random disturbance. This introduction of a random component is in line with random utility models (see, for example, Manski, 1977). One of many possible interpretations in this application is that randomness is due to the stochastic influence of the weather.

The systematic profit component is defined as:

$$v = p_m m - \Sigma_i p_i h_i - p_x x \tag{3.2}$$

where m is the production of maize per hectare; p_m is the market price of maize; p_i is the market price per hectare of the ith treatment; x is a vector of all inputs other than herbicides; and p_x is a vector of market prices of those inputs. When a farmer has made his choice of a certain treatment, all h_i values are zero, except for that of the selected treatment, which is unity.

It is also assumed that the production of maize can be described by a production function which relates inputs to production, i.e.:

$$m = L(h_1, \ldots, h_n, x) \tag{3.3}$$

However, since each herbicide treatment by assumption can be described by its particular combination of characteristics, the production function can also be written as:

$$m = M(z_1, \ldots, z_n, x) \tag{3.4}$$

We now also adopt the conventional hedonic hypothesis that the market price of a good is associated with a certain vector of characteristics. In the case of these herbicides, the hypothesis implies the existence of the following hedonic price function $P(.)$:

$$p_i = P(z_i) \qquad i = 1, \ldots, n \tag{3.5}$$

In general, this function does not reveal the structure of demand and supply, respectively, but is the combined result of demanders' and suppliers' behaviour. This means that a quite complex estimation procedure (or a set of very strong assumptions) is needed for obtaining information about the structure of demand and supply (see, for example, Epple, 1987, and Freeman, 1993, pp. 367–416, on this issue).

The discrete choice framework established by McFadden (1974) among others may offer a procedure for estimating the probability that a certain combination of characteristics is selected. Such a combination does not have to describe an existing treatment. There would therefore be a possibility of estimating the probability that a *new* type of treatment can be used.

The random profit model described above implies that the probability that the jth treatment is selected can be written as:

$$\text{Prob}\{h_j = 1\} = \text{Prob}\{\pi_j > \pi_i, \forall i \neq j\} \tag{3.6}$$

where π_j denotes the profits when the jth treatment is used, and π_i is the profits when the ith treatment is used.[4] This can also be expressed as $\text{Prob}\{h_j = 1\} = \text{Prob}\{\pi_j > \bar{\pi}\}$, where $\bar{\pi} = \max\{\pi_i, \forall i \neq j\}$. The random profit formulation implies that there are cumulative distribution functions of π_j and $\bar{\pi}$. These functions can be written as $F_{j1}(y) = \text{Prob}\{\pi_j \leq y\}$ and $F_{j2}(y) = \text{Prob}\{\bar{\pi} \leq y\} = \text{Prob}\{\pi_j \leq y, \forall i \neq j\}$, respectively.

It can be shown (see, for example, Cramer, 1991, pp. 50–1) that given an assumption that the random disturbances in equation (3.1) are independently and identically distributed (i.i.d.) according to a Weibull distribution[5], the probability that the jth treatment is selected can be written as:

$$\text{Prob}\{h_j = 1\} = \exp(\nu_j)/\Sigma_i \exp(\nu_i) \tag{3.7}$$

This standard result has been the basis for numerous discrete choice applications, and leads to what is usually called the conditional (or multinomial) logit model[6] see, for example, Greene, 1993, pp. 664–72). An estimation of such a model presupposes an empirical version of ν in terms of a linear combination of explanatory variables.

From equations (3.2), (3.4) and (3.5), we know that the systematic profits can be written as $\nu = p_m M(z_1, \ldots, z_n, x) - \Sigma_i P(z_i) h_i - p_x x$. This chapter focuses on only one of this equation's components, the hedonic price function. This constitutes, however, only a first step towards a full estimation of equation (3.7), and it is far from obvious what procedure is the most appropriate to follow for a full estimation. This issue still remains to be settled, and should be the subject of further research.

One possibility may be to have the profit function as a point of departure. This function relates prices of inputs and outputs to maximum profits (see, for example, Varian, 1992, pp. 40–5). This would suggest that an empirical version of ν can be defined in terms of the maize price, the implicit prices (shadow prices) of herbicide characteristics, i.e. the derivatives of the hedonic price function with respect to each characteristic, and other input prices. In other words, the basis for the empirical version of ν would be:

$$\nu = V(p_m, P_k, p_x) \tag{3.8}$$

Table 3.1. *Main treatment alernatives in Italian maize cultivation in 1987–1994*

Main treatment alternatives[a]	Period of use[b]	Main treatment alternatives[a] (cont.)	Period of use[b]
Atrazine	1987–1990	Metholachlor	1987–1994
Atrazine + alachlor	1987–1990	Metholachlor + terbutylazine	1987–1994
Atrazine + metholachlor	1987–1990	Metholachlor + pendimethalin	1987–1994
Atrazine + EPTC/butylate	1987–1988	Metholachlor + linuron	1987–1988,
Atrazine + pendimethalin	1987		1990–1994
Alachlor	1987–1994	Pendimethalin + linuron	1989–1994
Alachlor + pendimethalin	1987–1994	Primisulfuron	1991–1994
Alachlor + linuron	1987–1994	Primisulfuron + dicamba	1993–1994
Alachlor + terbutylazine	1989–1994		
		Rimsulfuron	1992–1994
		Rimsulfuron + dicamba	1993–1994
		Rimsulfuron + terbutylazine	1994

Notes:
[a] On average, the 19 treatments listed here constituted about 81% of total herbicide use.
[b] Some of the treatments were probably used, but unclassified, in some years. For example, this is probably true for metholachlor + linuron in 1989.

where $V(.)$ is the profit function, p_m is the market price of maize, $Pk = [\delta P(.)/\delta z_1, \ldots, \delta P(.)/\delta z_r]$ is a vector of implicit market prices of herbicide characteristics, and $p_x = [p_1, \ldots, p_s]$ is a vector of market prices of inputs. However, great care and quite detailed data are needed here. As shown by Brown and Rosen (1982), the implicit market prices may not provide any new information if restrictions on the functional form of the hedonic price function are not imposed, or if the implicit prices are not estimated from several different equations – for example, regional hedonic price functions.

Data[7]

Treatment price and characteristics

Thanks to close cooperation with a group of Italian scientists, the following market-oriented characteristics could be defined and objectively measured for the 19 main treatment alternatives (listed in Table 3.1) used in maize cultivation in Italy in 1987–1994: reliability (RELIA, measured as the GUS (groundwater ubiquity score), persistence (PERSI, measured as the soil half-life $(t_{1/2})$ value), action spectrum (ACTION, measured as the mean kill rate for a range of weed species) and toxicity (TOX, measured as the inverse of the LD_{50} value). ACTION is a dynamic characteristic in the sense that data for the mean kill rate of treatments during two different periods of time (1987–1988 and 1989–1994) were available. The values of the other characteristics are constant for each treatment during the whole period of 1987–1994 (see Sbriscia Fioretti *et al.*, 1995, for details). Note that the hedonic analysis will be confined to data on herbicides for pre-emergence treatment (i.e. the weeds are treated before the crops emerge) or for post-emergence treatment that does not follow a pre-emergence treatment. A minor part of the maize-cultivated area is subject to follow-up post-emergence treatments. The herbicides used for such a post-emergence treatment (typically dicamba, 2,4-D or MCPA) will not be considered here.

It was acknowledged that non-chemical properties such as advertising and treatment availability are likely to be of importance for farmers' treatment choices. In principle, it is also possible to measure these properties objectively by an analysis of advertisement frequencies and retailers' assortments. However, data that allow such a quantification are not available at present.

The choice of characteristics was influenced by the results of Infomark (1993) based on an interview survey of 319 maize-cultivating Italian farmers. The survey concerned, *inter alia*, the farmers' criteria for their selection of herbicide treatments. The most important criteria was 'experience with the product' (69% of the interviewed farmers reported this criterion). Other important criteria include 'low cost' (55%), 'absence of residues in soil' (54%), 'selectivity in relation to corn' (42%), 'toxicological class' (41%) and 'wide range of action' (38%). The least important criteria turned out to be 'specific action on dicotyledons' (7%) and 'company's assistance' (15%) (see also Sbriscia Fioretti *et al.*, 1995, Table 5).

The importance of toxicity and a wide action spectrum is reflected by TOX and ACTION, respectively. Absence of residues, which implies that a subsequent crop to be cultivated is not damaged by toxic residues, is negatively correlated with PERSI. However, PERSI is an ambiguous characteristic, since a greater persistence is related to a prolonged treatment efficacy. In the interview survey, the latter persistence criterion was found to be less important than the 'absence of residues' criterion (29% vs. 54%). RELIA is defined by Sbriscia Fioretti *et al.* (1995, p. 13) as 'capacity of a herbicide to maintain its biological action with the variation of pedo-climatic conditions and growth stage of the weeds'. This capacity is likely to be positively correlated with the GUS index, though this index is usually employed for the classification of herbicides into leachers (to groundwater; high GUS value) and non-leachers (low GUS value) (see Gustafson, 1989 and Vighi and Di Guardo, 1995). Among the criteria identified by Infomark [1993], 'ease of use' (34%) is probably positively correlated with RELIA, since a low sensitivity to varying agronomic and environmental conditions makes a treatment easier to use. Though selectivity seems to be an important criterion (42%), a satisfactory way to measure this characteristic was unfortunately not found.

The cost criterion is of course related to the treatment price (PRICE), i.e. the dependent variable of the hedonic price function. PRICE comprises information on both the herbicide price per kilogram and the dose in kilograms per hectare, i.e. PRICE is measured in ITL 1 000 (kITL) per hectare. Annual price data were available for each treatment but, due to the mainly constant nature of the characteristics, PRICE was defined as the mean price of each 'kill rate period' (1987–1988 and 1989–1994) for each treatment. See Table 3.2 for a statistical description of PRICE and the treatment characteristics.

The observations are plotted in Figures 3.1*a–d*. The figures also include simple two-variable regression lines. In correspondence with the partial correlations in Table 3.1, the regressions suggest that there is a positive relationship between ACTION and PRICE, no relationship at all between TOX and PRICE, whereas the relationships between RELIA and PRICE and between PERSI and PRICE seem to be negative.

Estimation of a hedonic price function

A basic model

A basic version of the hedonic price function for maize herbicides $P(z_i)$ was estimated by regressing PRICE on the four characteristics RELIA, PERSI, ACTION and TOX. A standard problem in such an estimation is that all characteristics of importance for herbicides price are not likely to be included. In this study, the most obvious examples are characteristics that

Table 3.2. *Statistical description of the variables used in the hedonic price function*

Variable	Mean	Median	Range	SD
PRICE: treatment price per hecture in kITL and 1993 prices (dependent variable)	70.210	68.750	14.050–140.100	28.212
RELIA: GUS value	2.15	2.40	0.24–3.20	0.888
PERSI: $t_{\frac{1}{2}}$	38	37	12–80	17.619
ACTION: mean kill rate	0.764	0.749	0.575–0.920	0.077
TOX: $1/LD_{50}$	0.2480	0.2464	0.0500–0.3897	0.101

Correlation matrix

	PRICE	RELIA	PERSI	ACTION	TOX
PRICE	1				
RELIA	−0.375	1			
PERSI	−0.452	0.153	1		
ACTION	0.765	−0.066	−0.505	1	
TOX	0.071	−0.415	−0.325	−0.144	1

Notes:
Number of observations: 29. Number of treatments: 19. SD, standard deviation.

The only dynamic characteristic ACTION was defined for two periods (1987–1988 and 1989–1994). Ten of the 19 treatments were available in both periods. Since annual price data were available, PRICE is defined as the mean price of thise two periods for each treatment.

For five post-emergence treatments, the reliability index of Sbriscia Fioretti *et al.* (1995, Table 10) was transformed to GUS values by the use of the categories 'low reliable', 'fairly reliable' and 'very reliable'. These categories were defined by Sbriscia Fioretti *et al.* for both the reliability index and the GUS index.

consider selectivity and the non-chemical properties mentioned in the previous section. Some relevant explanatory variables correlated with the ones included in the regression are thus likely to have been omitted. This causes an estimation bias of unknown magnitude. Another standard problem is the choice of functional form for the hedonic price function. Since theory gives no advice concerning the functional form, it is generally of importance to compare the estimation results of several different specifications. In the following, the results of three different linear

(*a*)

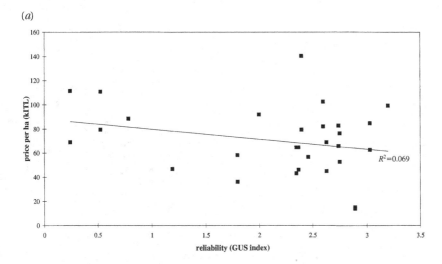

(*b*)

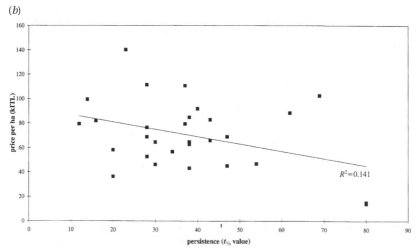

Fig. 3.1. (*a*) Plot of observations: RELIA against PRICE.
(*b*) Plot of observations: PERSI against PRICE.

specifications will be presented: all variables untransformed (A, 'linear');
natural logarithms of the explanatory variables, but an untransformed
dependent variable (B, 'semi-log'); and natural logarithms of all variables
(C, 'log-log'). A more refined analysis could use Box–Cox transformations
in order to find a specification that can be regarded as optimal (see, for
example, Cropper *et al.*, 1988 and Greene, 1993, pp. 329–35).

Ordinary least squares (OLS) estimation results for specifications A, B

(c)

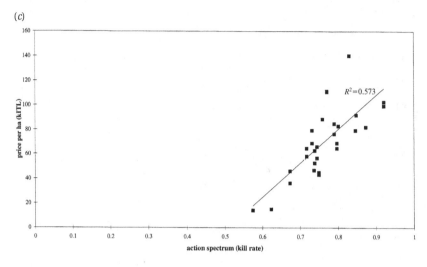

(d)

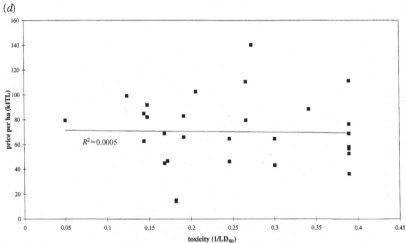

Fig. 3.1. (cont.) (c) Plot of observations: ACTION against PRICE.
(d) Plot of observations: TOX against PRICE.

and C are presented in Table 3.3. Similar results are obtained for all three specifications. The lowest F value is 14.09, which implies that the hypothesis that all coefficients are equal to zero can be rejected at a very low level of significance (<0.01). This is also reflected by the rather substantial adjusted R^2 values (0.65–0.80). The low values of the Breusch–Pagan X^2 statistic suggest that a null hypothesis of presence of homoscedasticity cannot be rejected. The condition numbers indicate that multicollinearity may be

Table 3.3. *OLS estimation results for the hedonic price functions:*
A. Linear: $\text{PRICE}_i = \beta_0 + \beta_1\text{RELIA}_i + \beta_2\text{PERSI}_i + \beta_3\text{ACTION}_i + \beta_0\text{TOX}_i$
B. Semi-log: $\text{PRICE}_i = \beta_0 + \beta_1\ln(\text{RELIA}_i) + \beta_2\ln(\text{PERSI}_i) + \beta_3\ln(\text{ACTION}_i) + \beta_4\ln(\text{TOX}_i)$
C. Log-log: $\ln(\text{PRICE}_i) = \beta_0 + \beta_1\ln(\text{RELIA}_i) + \beta_2\ln(\text{PERSI}_i) + \beta_3\ln(\text{ACTION}_i) + \beta_4\ln(\text{TOX}_i)$
$i = 1, \ldots, 29$

Coefficient	Estimate (t value)		
	A	B	C
β_0 (Constant)	−132.07*	146.58**	6.0242**
	(−2.771)	(5.556)	(16.034)
β_1 (RELIA)	−10.431*	−12.237*	−0.19799**
	(−2.606)	(−2.673)	(−3.037)
β_2 (PERSI)	−0.056315	−0.18031	−0.093279
	(−0.274)	(−0.025)	(−0.924)
β_3 (ACTION)	292.17**	223.70**	4.3409**
	(6.045)	(6.548)	(8.923)
β_4 (TOX)	14.771	4.7779	0.16023
	(0.385)	(0.643)	(1.515)
Adjusted R^2	0.652	0.667	0.800
$F(4, 24)$	14.09	15.12	29.03
Breusch–Pagan $\chi^2(4)$	2.74	3.76	0.48
Condition number	41.5	24.0	24.0

Notes:
** and * represent coefficients significantly different from zero at a 1% and 5% level of significance, respectively.

Note that the presence of partly grouped data (mean prices for two periods) may justify the use of weighted least squares instead of OLS; see Greene (1993, pp. 277–9). Such an estimation procedure gives results very similar to those reported in this table. These results are available from the author upon request.

present in specification A. This number is much lower in specifications B and C. The coefficient estimates for ACTION are positive in all three specifications. Moreover, the null hypothesis that the coefficient is equal to zero is rejected at a very low level of significance (<0.01). The results for RELIA are unambiguous: the coefficient estimates are all negative, and the null hypothesis can be rejected at a level of significance <0.02. The coefficient estimates for PERSI are also negative, but there is no statistical significance.[8] Insignificance is also the case for the coefficients for TOX, whose estimates have positive signs.

Interestingly, the results reported in Table 2.3 correspond fairly well to the findings of Beach and Carlson (1993). They also obtained estimated hedonic price functions which had large F values, significant and positive coefficients of the kill rate characteristics, and significant and negative coefficients of the GUS characteristic. The coefficient of their $t_{1/2}$ characteristic became significant and negative. Their $1/LD_{50}$ characteristic had negative and mostly insignificant coefficients. These results will be discussed more later (pp. 67–8).

Sensitivity analysis

An analysis of unusual observations was undertaken. When one observation with a standardised residual >2 (atrazine + EPTC/butylate) and three observations with leverages >2 × (5/29) = 0.345 (atrazine, rimsulfuron and rimsulfuron + terbutylazine) are excluded, the only large changes concern the coefficients of PERSI and TOX. The sign of the estimates of the coefficient of PERSI becomes positive in specifications B and C, and the coefficient estimate of TOX becomes negative in all three specifications. However, their statistical insignificance does not change.

The results in Table 3.3 and Figure 3.1c suggest that ACTION is the key variable in the hedonic price function. It could therefore be of interest to study the consequences of excluding this variable. Not surprisingly, OLS regressions without ACTION result in a considerably lower explanatory power: adjusted R^2 values from 0.11 (specification B) to 0.17 (specification C). The coefficients of RELIA and PERSI remain negative, and are now significantly different from zero at a level <0.1. The coefficients of TOX become negative but are still insignificant.

It was also checked whether any differences would arise when separate hedonic price functions are estimated for each of the eight years in the period of 1987–1994 (and for each of the specifications A, B and C). It is possible to develop this type of quite informal analysis into a formal analysis of pooling. A complication is that, in any event, the number of observations is small. We stick to the informal analysis here in order to obtain some rough indications of potential coefficient instability over time. Figures 3.2a and b show the results for these annual hedonic price

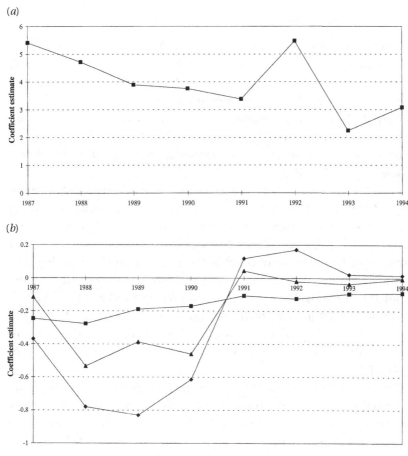

Fig. 3.2. (*a*) Specification C estimated for each year. ■, ACTION.
(*b*) Specification C estimated for each year. ■, RELIA; ◆, PERSI; ▲, TOX.

functions when specification C is used. The results for the other two specifications are similar. Figure 3.2*a* presents the coefficient estimates for ACTION. These are positive without exception and mostly significant. The coefficient estimates for RELIA are negative in all years in all specifications, but significantly so only in some cases. A small number of observations is of importance here, but note from Figure 3.2*b* that there seems to be a slight tendency over time towards less negative coefficient estimates. As regards PERSI and TOX, most of the annual coefficient estimates were negative. There are, however, only very few cases of statistical significance for these coefficients.

The supply side

An obvious shortcoming with the analysis so far is that the supply side has not been considered explicitly. It is possible that it has been taken into account implicitly, since the four studied characteristics may be determinants not only for farmers' treatment choices, but also for producers' supply decisions. This complex issue is discussed in the last section. However, we now turn to study how the estimated hedonic price function is influenced by the introduction of some variables that are related explicitly to the supply of herbicides.

First, the legislative actions against atrazine can be considered in a more refined way. Regional legal restrictions against treatments containing atrazine were introduced on an increasing scale during the period 1986–1989. In 1990, a national ban was placed on the sale and use of atrazine (cf. Faure, 1994), and treatments containing atrazine are therefore not included in the data from 1991, see Table 3.1. In order to take the gradual restrictions also into account, the variable RESTR is defined as the maize-cultivated area for which restrictions were active divided by the total area cultivated with maize. However, for all treatments that did not contain atrazine, RESTR takes a zero value. See Table 3.4 for a statistical description of RESTR.

Secondly, Table 3.1 shows that the herbicide market was dynamic during the studied period, in the sense that several new treatment alternatives were introduced. In order to account for the potential special pricing policy of producers who release a new product, the variable INTRO describes how new the treatment is to the Italian herbicide market by assigning the year of introduction to each treatment, see Table 3.4 for details. At least two counteracting forces are conceivable here. A new product may be patented, and Beach and Carlson (1993) found a positive relationship between the existence of a patent and herbicide expenditures. On the other hand, a low price policy may be of help in the successful introduction of a new product.

Finally, INTRO will be supplemented by another variable that describes the market structure. In order to obtain at least a rough indication of producers' possibly different pricing policy for treatments that have a leading market position, the variable LEADERS is introduced. It is defined quite arbitrarily as a dummy variable that takes the value of unity for the two principal treatments in each period, and zero otherwise (see Table 3.4 for details).

Table 3.4. *Statistical description of variables related to the supply side*

Variable	Mean	Median	Range	SD
RESTR. For treatments containing atrazine: the maize-cultivated area in northern Italy for which restrictions were active divided by the total area cultivated with maize in northern Italy for the period in question. For all other treatments, RESTR takes a zero value	0.068	0	0–0.445	0.13912
INTRO. The year of introduction for a treatment (1987 or earlier = 0, 1988 = 1, ..., 1994 = 7)	1.1034	0	0–7	2.1923
LEADERS. A dummy variable which takes the value of unity for the two principal treatments for the period in question, and zero for all other treatments. 'The two principal treatments' are defined as the treatments which accounted for the two largest proportions of the total treated maize-cultivated area in northern Italy	0.13793	0	0/1	0.35093

Correlation matrix

	PRICE	RELIA	PERSI	ACTION	TOX	RESTR	INTRO	LEADERS
PRICE	1							
RELIA	−0.263	1						
PERSI	−0.375	0.078	1					
ACTION	0.757	0.117	−0.382	1				
TOX	−0.022	−0.463	−0.189	−0.293	1			
RESTR	−0.411	0.198	0.442	−0.330	−0.098	1		
INTRO	0.304	0.212	−0.145	0.706	−0.464	−0.255	1	
LEADERS	0.004	−0.060	0.000	0.023	0.082	−0.104	−0.205	1

Note:
SD, standard deviation.

Table 3.5. *OLS estimation results for the hedonic price functions:*
1. As specification C in Table 3, but RESTR included
2. As specification C in Table 3, but RESTR and INTRO included
3. As specification C in Table 3, but RESTR, INTRO and LEADERS included

Coefficient	Estimate (t value)		
	1	2	3
β_0 (Constant)	5.8030**	5.9990**	6.0142**
	(14.903)	(20.011)	(20.657)
β_1 (RELIA)	−0.17781*	−0.16228**	−0.16602**
	(−2.760)	(−3.303)	(−3.477)
β_2 (PERSI)	−0.042577	−0.052214	−0.036577
	(−0.414)	(−0.667)	(−0.477)
β_3 (ACTION)	4.1446**	5.1757**	5.3332**
	(8.509)	(11.658)	(12.039)
β_4 (TOX)	0.15036	−0.00017151	−0.0070090
	(1.465)	(−0.002)	(−0.084)
β_5 (RESTR)	−0.56401	−0.69355*	−0.75060**
	(−1.601)	(−2.572)	(−2.839)
β_6 (INTRO)	—	−0.093850**	−0.10395**
		(−4.214)	(−4.600)
β_7 (LEADERS)	—	—	−0.14627
			(−1.533)
Adjusted R^2	0.812	0.891	0.898
F	$F(5,23)=25.25$	$F(6,22)=39.3$	$F(7,21)=36.1$
Breusch–Pagan χ^2	$\chi^2(5)=0.70$	$\chi^2(6)=4.91$	$\chi^2(7)=10.25$
Condition number	26.5	27.2	27.6

Note:
** and * represent coefficients significantly different from zero at a 1% and 5% level of significance, respectively.

Table 3.5 shows the consequences of adding the variables RESTR, INTRO and LEADERS to specification C. Similar results in terms of coefficient signs, but a somewhat weaker statistical significance, are obtained if specifications A or B is used instead. The three variables are added sequentially: first the most obvious variable in terms of supply restrictions (RESTR), then INTRO, and finally the most arbitrarily defined one, LEADERS. The results indicate that all these variables have a negative impact on the market price. The coefficients of RESTR and INTRO are even significantly different from zero when both of these variables are included in the hedonic price function. For INTRO at least this is a rather surprising finding, since the correlation matrix in Table 3.4 gives a positive partial correlation between INTRO and PRICE (0.304). However, the substantial positive correlation between INTRO and ACTION (0.706), indicating that new treatments tend to have a relatively high kill rate, may confuse the issue. If ACTION is excluded from the regressions, the condition number is reduced, and the coefficient of INTRO changes its sign to positive in all three specifications. The t value is, however, only slightly greater than unity. Also the coefficient of LEADERS changes to positive, but the t value is even lower in this case. The coefficients of RELIA, PERSI and RESTR are considerably more robust, though much of the explanatory power disappears when ACTION is excluded.

To summarise: an increase in restrictions (as measured by RESTR) seems to be related to a reduction in market prices. Atrazine was, before the introduction of restrictions, a low price treatment alternative. When the positive correlation between INTRO and ACTION is taken into account, the results indicate that new treatment alternatives have a relatively high market price. This may reflect patent status. The attempt to account for market leaders by LEADERS gave no clear results at all. Finally, it should be emphasised that when these explicit supply variables are included in the analysis, the other herbicide characteristics (RELIA, PERSI, ACTION and TOX) may still be determinants for both demand and supply decisions.

Discussion

The most robust results in the empirical analysis of the preceding section are the positive relationship between kill rate (ACTION) and herbicide price, the high explanatory power of ACTION in the regressions, and the negative impact of reliability (measured as the GUS value) on herbicide price. The following discussion therefore focuses on these two variables.

It was noted above that the results for ACTION and RELIA correspond well to those reported by Beach and Carlson (1993). This is an encouraging fact, but this does not mean that the results are easy to interpret. Recall from the description of the model that it is typically difficult to interpret an estimated hedonic price function in terms of demand and supply. However, the case of ACTION is hardly problematic. Beach and Carlson suggested that the kill rate is an unambiguously productive characteristic in maize cultivation and should thus influence the treatment price positively. A high kill rate is valuable to farmers and it is probably costly to design herbicides that achieve a relatively high kill rate for an appropriate set of weed species.

It is more difficult to find a straightforward interpretation of the results for the GUS index (RELIA). Beach and Carlson regarded this variable solely as an index for water quality. They therefore argued that the negative sign and significance of this characteristic in a hedonic price function indicates that water quality is an appreciated component in farmers' selection of treatment. In this study, the GUS index is used as a proxy for reliability of a herbicide, i.e. a productive characteristic. However, Beach and Carlson's finding of a negative and significant coefficient is confirmed. This may imply that water quality issues also are appreciated by Italian farmers. This of course presupposes that information on the relationship between herbicide characteristics and the appearance of herbicides in groundwater is widely known by the farmers. This can be questioned, although the restrictions against the use of atrazine are likely to have brought groundwater quality issues to the forefront of publicity. Any concern for groundwater quality among farmers is not evident from the results of the Infomark (1993) survey, but it is not known whether this issue was considered explicitly by the interviewers.

However, there may also be other explanations for the negative relationship between RELIA and PRICE. One quite appealing speculation, which takes both the demand side and the supply side into account, is the following. It may very well be the case that the GUS index works as a proxy for reliability in the sense that a high reliability is obtained 'free of charge' for herbicides characterised by a high GUS value. However, producers and farmers probably want a satisfactory reliability also for treatments with a low GUS value. This has to be accomplished by other means, which may imply an increase in production costs and, as a consequence, in prices.

Unfortunately, even if this or some other speculation is found to be reasonable, it cannot be turned into fact without further theoretical and empirical analysis. What we do know, however, is that herbicides that are more likely to contaminate groundwater (a relatively high GUS value) have a lower market price than other herbicides, other things being equal. This is a significant result found by Beach and Carlson for the USA as well as in this study. Hence, herbicide prices do not provide any incentive for the use of non-leaching herbicides, which may have serious consequences in northern Italy, where protection of groundwater resources is important for, *inter alia*, the preservation of a good drinking-water quality. As was noted above, groundwater is the dominant source of drinking water in northern Italy. It is in this context interesting to note that the inhabitants of Milan at least seem to place a rather high monetary value on preserving drinking-water quality (see Press and Söderqvist, 1996). Reducing the risk of herbicide contamination of groundwater by increasing the price of leaching substances is likely to be an important contribution to quality preservation.[9]

Finally, the hedonic analysis in this chapter may provide a basis for a more advanced analysis of farmers' selection of treatment alternatives and the prediction of future herbicide use. This presupposes, however, a successful combination of the hedonic model and the treatment selection model sketched earlier. As was noted on p. 53, the way to accomplish such a combination has still to be studied carefully. Regional hedonic price functions may be necessary for estimations, and individual date on farmers' treatment choices are desirable.

Notes

1 I am grateful for help and constructive comments from Andrea Bassi, Lars Bergman, Per-Olov Johansson, Robin Mason, Marc Nerlove, Carolina Sbriscia Fioretti, Timothy Swanson, Marco Vighi, Giuseppe Zanin and other members of Working Group 4 of the European Science Foundation Programme on Environmental Damage and Its Assessment. I am also indebted to the Beijer International Institute of Ecological Economics for hospitality. The usual disclaimer applies. Financial support from the European Science Foundation and the Swedish Council for Planning and Coordination in Research is gratefully acknowledged.

2 See, for example, Freeman (1993) on this method.

3 See, for example, Griliches (1971; 1988) on the modern approach to hedonic price indexes.

4 A possibly fruitful alternative specification of $\text{Prob}\{h_j = 1\}$ could be based on profit differences $\Delta\pi_{ji} \equiv \pi_j - \pi_i$ (cf. the utility differences of Hanemann, 1984). In this case, $\text{Prob}\{h_j = 1\} = \text{Prob}\{\Delta\pi_{ja} > \Delta\pi_{ia}, \forall i \neq j\}$, where $\Delta\pi_{ia} \equiv \pi_i - \pi_a$. π_a may represent some benchmark profits.

5 A Weibull distribution has the distribution function $F(y) = \exp - \exp(- y)]$. It is sometimes also called a Gumbel distribution or a type I extreme value distribution in standard form.

6 Note that a conditional logit model is characterized by what is called the indepen-

dence-of-irrelevant-alternatives (IIA) property. This property implies that the conditional logit model is inappropriate whenever different choice alternatives are very close substitutes (the standard example is red buses and blue buses for commuters) (see, for example, Kennedy, 1992, p. 244, and Greene, 1993, pp. 670–2). While the treatment alternatives in this study involve deliberate differences, this property may still constitute a problem. An obvious alternative is a conditional probit model. The estimation of such a model is complex due to computational difficulties, but may be accomplished by simulation techniques (Nerlove and Schuermann, 1995).

7 Data were collected by Dr Carolina Sbriscia Fioretti, Dr Paolo Ferrario and Professor Marco Vighi, all at the University of Milan, and Professor Giuseppe Zanin, University of Padua.

8 'No statistical significance' and 'insignificance' in this chapter refer to situations when a null hypothesis can be rejected at a level of significance >0.10.

9 Mason and Swanson (1993) described the principles for a protection policy that combines an accumulation charge with a rent on the remaining stock of water quality.

References

Beach, E. D. and Carlson, G. A. (1993). A hedonic analysis of herbicides: do user safety and water quality matter? *American Journal of Agricultural Economics*, **75**, 612–23.

Bergman, L. and Pugh, D. M. (eds.) (1994). *Environmental Toxicology, Economics and Institutions: The Atrazine Case Study*. Kluwer Academic Publishers, Dordrecht.

Brown, J. N. and Rosen, H. S. (1982). On the estimation of structural hedonic price models. *Econometrica*, **50**, 765–8.

Court, A. T. (1939). Hedonic price indexes with automative examples. In *The Dynamics of Automobile Demand*, pp. 99–117. General Motors, New York.

Cramer, J. S. (1991). *The Logit Model*. Hodder & Stoughton Ltd, Sevenoaks.

Cropper, M. L., Deck, L. B. and McConnell, K. E. (1988). On the choice of functional form for hedonic price functions. *Review of Economics and Statistics*, **70**, 668–75.

Epple, D. (1987). Hedonic prices and implicit markets: estimating demand and supply functions for differentiated products. *Journal of Political Economy*, **95**, 59–80.

Faure, M. G. (1994). The EC Directive on drinking water: institutional aspects. In *Environmental Toxicology, Economics and Institutions: The Atrazine Case Study*, ed. L. Bergman and D. M. Pugh, pp. 39–87. Kluwer Academic Publishers, Dordrecht.

Freeman, A. M., III (1993). *The Measurement of Environmental and Resource Values: Theory and Methods*. Resources for the Future, Washington, DC.

Greene, W. H. (1991). *Limdep. User's Manual and Reference Guide*. Econometric Software, Bellport, NY.

Greene, W. H. (1993). *Econometric Analysis*, second edition. Macmillan, New York.

Griliches, Z. (1971). Introduction: hedonic price indexes revisited. In *Price Indexes and Quality Change: Studies in New Methods of Measurement*, ed. Z. Griliches, pp. 3–15. Harvard University Press, Cambridge, MA.

Griliches, Z. (1988). *Technology, Education, and Productivity*. Basil Blackwell, New York.

Gustafson, D. I. (1989). Groundwater ubiquity score: a simple method for assessing pesticide leachability. *Environmental Toxicology and Chemistry*, **8**, 339–57.

Hanemann, W. M. (1984). Welfare evaluations in contingent valuation experiments with discrete responses. *American Journal of Agricultural Economics*, **66**, 332–41.

Infomark (1993). Market survey, courtesy of Du Pont de Nemours and Co., Inc.

Kennedy, P. (1992). *A Guide to Econometrics*, third edition. Basil Blackwell, Oxford.

Lancaster, K. J. (1966). A new approach to consumer theory. *Journal of Political Economy*, **74**, 132–57.

Mason, R. and Swanson, T. M. (1993). Regulating chemical waste: addressing manufacturer incentives. Faculty of Economics and Politics, University of Cambridge. (Mimeo)

McFadden, D. (1974). Conditional logit analysis of qualitative choice behavior. In *Frontiers in Econometrics*, ed. P. Zarembka, pp. 105–42. Academic Press, New York.

Manski, C. F. (1977). The structure of random utility models. *Theory and Decision*, **8**, 229–54.

Nerlove, M. and Schuermann, T. (1995). Expectations: are they rational, adaptive or naive? An essay in simulation-based interference. In *Advances in Econometrics and Quantitative Economics: Essays in Honor of Professor C. R. Rao*, ed. G. S. Maddala, P. C. B. Phillips and T. N. Srinivasan, pp. 354–81. Basil Blackwell, Oxford.

Press, J. and Söderqvist, T. (1996). On estimating the benefits of groundwater protection: a contingent valuation study in Milan. Fondazione Eni Enrico Mattei, Milan, and Beijer International Institute of Ecological Economics, Stockholm. (Mimeo)

Rosen, S. (1974). Hedonic prices and implicit markets: product differentiation in pure competition. *Journal of Political Economy*, **82**, 34–55.

Sbriscia Fioretti, C., Zanin, G., Ferrario, P. and Vighi, M. (1995). Weed control on maize in Italy: agronomic and environmental inputs for an economic analysis. Istituto di Entomologia Agraria, Università degli Studi di Milano. (Mimeo)

Varian, H. R. (1992). *Microeconomic Analysis*, third edition. W. W. Norton & Company, New York.

Vighi, M. and Di Guardo, A. (1995). Predictive approaches for the evaluation of pesticide exposure. In *Pesticide Risk in Groundwter*, ed. M. Vighi and E. Funari, pp. 73–100. Lewis Publishers, Boca Raton, FL.

Zanin, G., Sattin, M. and Berti, A. (1995). Innovative strategies in weed control. *Pesticide Risk in Groundwater*, ed. M. Vighi and E. Funari, pp. 73–100. Lewis Publishers, Boca Raton, FL.

Part II

Estimating the costs of chemical accumulation

4 Environmental toxicology: the background for risk assessment

Marco Vighi, Richard Lloyd and Carolina Sbriscia Fioretti

Introduction

The potential harmfulness of a chemical in the environment depends on several major properties: toxicity to organisms, bioaccumulation in tissues, persistence, mobility and distribution patterns in environmental compartments. Historically, the focus of ecotoxicology was on the first property, toxicity, and techniques were devised to enable maximum concentrations of no harmful effect to be established, usually using responses measured in whole organisms. These concentrations then formed the basis of environmental quality standards.

For many years, these studies played a vital role in the control of the major causes of chemical pollution from industry. However, in the past three decades, there has been an increasing awareness of the special importance of persistent chemicals, which can be transported and detected far from their original source; this problem was highlighted by the discovery of widespread environmental contamination by DDT and later by polychlorinated biphenyls (PCBs). Organisms over a wide area could be exposed to low concentrations of such substances for a long period of time. Ecotoxicologists then began to search for very sensitive biological responses in order to detect the effect, if any, of these low concentrations.

Considerable attention was focused on organismal effects at the cellular or subcellular level, where they first become apparent. These were often found at exposure concentrations lower than those predicted to be safe from tests on whole organisms. The ensuing scientific and political debate served only to cloud the validity of existing environmental quality standards, and indeed shed doubt on the value of ecotoxicological data in pollution prevention and control.

At the same time, the definition of pollution was changing. The original definition (Lloyd, 1991) referred to damage caused to environmental resources (e.g. fisheries) and was thus based on *effects*, but this was superseded by an alternative definition based on the *presence* of synthetic chemicals or chemicals derived from human activity. This latter definition focused more on the property of fate and persistence than on toxicity.

To some extent, this change in definition followed from the development of extremely sensitive methods of chemical analysis. These enabled very low concentrations of chemicals to be measured in environmental compartments where they had previously been presumed to be absent. This fed a general concern that pollution was becoming more widespread; the fact that these minute quantities were likely to cause no harmful effects to human beings or other organisms was overridden by a public perception that truly safe concentrations of these chemicals could not be established.

It is not surprising, therefore, that in the last decade considerable attention has been given by ecotoxicologists to measuring and predicting the movement and persistence of chemicals in and among environmental compartments. Models have been constructed which have considerable practical value. To some extent, this development followed from the procedure used to predict the potential environmental hazard of new chemicals, in which the predicted environmental concentration (PEC) was compared with the 'no effect' concentration (NEC). However, it is becoming necessary to compare the PEC with the limit of analytical detection, which requires a greater degree of accuracy of prediction. On the other hand, the rigorous and precise quantification of the NEC is not a scientifically attainable result. As described in more detail in the following paragraphs, scientific evidence can be given only for a 'no *observed* effect' concentration (NOEC) in relation to the extent of environmental information available, and this can be only an approximation of the real NEC.

Therefore, more or less precautionary Environmental Quality Objectives could be developed, not only on toxicological bases, but also with respect to what has to be protected and how high the level of protection must be.

The main objective of ecotoxicology is therefore to produce suitable tools for the setting up of Environmental Quality Objectives and for the evaluation and prediction of environmental concentrations, in order to assess the risk for natural populations (including human), potentially exposed to environmental contaminants. In this chapter the essential of ecotoxicological background needed to understand the procedures for effect and exposure assessment in order to evaluate environmental risk will be described briefly.

The assessment of effects

Toxicology and ecotoxicology

Toxicology can be defined as: 'the science that defines limits of safety of chemical agents' (Casarett and Doull, 1975). This definition can be applied also to that part of ecotoxicology dealing with the assessment of effects of chemical agents on living organisms and ecosystems. Nevertheless, there are some substantial differences between toxicology and ecotoxicology in the objective to be attained and in the methods used to attain it.

The results of experimental toxicology are produced using a number of test animals, generally mammals (rats, rabbits, monkeys, etc.), and data are used to produce limits of safety for the primary object of this approach – human beings. The object of ecotoxicology, conversely, is ideally the protection of the whole biosphere, comprising millions of very different species of living organisms, organised in populations, communities, and ecosystems regulated by complex interactions. Moreover, experimental toxicity testing should be limited to a few test organisms, assumed to be representative of the different levels of biological and ecological organisation. The results must then be extrapolated in order to produce quality objectives able to protect the global environment.

The aim of human toxicology is the protection of individuals. In *natural* ecosystems, however, the death of a large number of individuals of different species is a normal process resulting from natural selection and the fight for survival. The main aim of ecotoxicology therefore is not to protect individuals but to preserve the structures and functions of the ecosystem. How many individuals could be lost without significant damage to the ecosystem? This is another crucial question in defining Environmental Quality Objectives.

Toxicological and ecotoxicological testing

Experimental approaches for the assessment of the effects of potentially dangerous chemicals can be developed at different levels of complexity, which take into account either the organisation of the studied target (single species, population, community, ecosystem), or the type of end-point (short- or long-term mortality, chronic or subchronic responses, reproductive impairment, etc.). The need for a compromise between 'ecological realism' and the simplicity of the procedure and interpretation of results is

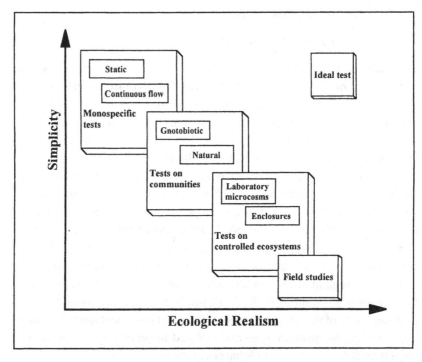

Fig. 4.1. The relationship between ecological realism and simplicity of procedure and interpretation for different ecotoxicological testing approaches (modified from Blank *et al.*, 1978).

a long-standing debate (Blank *et al.*, 1978) and the problem of the development of an 'ideal testing procedure' is still unresolved (Figure 4.1). Advantages and limitations of different testing approaches and the possibilities for their application are shown in Tables 4.1 and 4.2.

For the comparative screening of chemicals, and in particular for legal purposes, there is a need for standardised and reproducible methods, and present community and ecosystem approaches do not show these features. Therefore legal testing procedures are based mainly on single-species testing, even if relatively complex end-points are often required (chronic effects, fertility studies, etc.).

For practical purposes, the results of either a simple toxicological test or a complex toxicological procedure must be expressed quantitatively in a form readily understandable and easily comparable with other data. Three quantitative indices are most commonly used in toxicology:

Table 4.1. *Possibility of application and informative content of different ecotoxicological testing approaches*

	Monospecific tests	Community tests	Controlled ecosystem	Field studies
Screening toxicity of chemical substances	+	±	−	−
Monitoring toxicity of waste emissions	±	±	−	−
Monitoring environmental quality	±	−	−	+
Assessing sensitivity of natural environment to chemical substances	+	±	±	−
Research to establish environmental quality criteria	±	±	+	+

Note:
The symbol + means a positive evaluation; − means a negative evaluation; ± means medium validity or a conflicting situation.

Table 4.2. *Value and limitations of different ecotoxicological testing approaches*

	Monospecific tests	Community tests	Controlled ecosystem	Field studies
Practicability	+	±	−	−
Reproducibility	+	±	−	−
Flexibility	+	+	±	−
Informative content	±	±	+	+
Ease of interpretation	+	+	±	±
Predictive value	±	±	±	±

Note:
The symbol + means a positive evaluation; − means a negative evaluation; ± means medium validity or a conflicting situation.

LD_{50} *(lethal dose 50%):* It represents the dose of a chemical substance that will produce death in 50% of the experimental
animals. LD_{50} is a statistically obtained virtual value, calculated
using the results of an experimental test. LC_{50} (lethal concentration 50%) is similar and is used if the toxic testing procedure is not
via oral administration but via other exposure routes (water, air).
If the end-point is not death but another biological effect, the
concepts of ED_{50} or EC_{50} (effective dose or effective concentration
50%) are used. If time is an important component of exposure, it
must be indicated. For example LC_{50} 24h is the concentration that
will produce death of 50% of living organisms in 24 hours.

LT_{50} *(lethal time 50%):* Another approach consists of measuring the
time needed for the death of 50% of treated organisms. This is
obtained by keeping the incidence of the effect constant and
experimentally observing the time needed to reach this effect as
a function of concentration or dose.

NOEL (no observed effect level): This represents the highest
concentration or dose that does not show any experimental evidence of an effect on test organisms. Technically, a NOEL could
be applied also to a short-term acute test. Nevertheless, in
toxicological praxis a NOEL should be derived from a relatively
wide data set, including long-term chronic experiments. At the
other end of the real toxicity threshold lies the LOEL (lowest
observed effect level).

Limits of safety for human beings and the ecosystem

The setting of environmental quality objectives

In EEC terminology an environmental quality objective expresses the
maximum pollution levels to be satisfied in the environment in order to safe-
guard an effective protection of the ecosystem by 'taking into account the so-
called *zero-effect evaluation*' (EEC, 1973). The term is comparable with the
water quality criteria applied by the US Environmental Protection Agency
(USEPA, 1987) and by the European Inland Fisheries Advisory Committee
(Lloyd and Calamari, 1987), and also with the concepts of environmental
concern level of the Organisation for Economic Co-operation and
Development (OECD, 1992) and maximum permissible concentration

(MPC) from the Dutch practice (Sloof *et al.*, 1986), all of which tend to define levels of protection at which unacceptable, adverse (ecotoxicological) effects on the ecosystems are not likely to occur.

A possible definition of quality objective has been proposed in more detail for the aquatic environment by the Scientific Advisory Committee on Ecotoxicity and Environment (CSTEE) of the European Union. According to this definition (CSTEE / EEC, 1987) a Water Quality Objective (WQO):

> should be such as to permit all stages in the life of aquatic organisms to be successfully completed;
> should not produce conditions that cause these organisms to avoid parts of the habitat where they would normally be present;
> should not give rise to the accumulation of substances that can be harmful to the biota (including man) whether via the food chain or otherwise;
> and should not produce conditions that alter functioning of the ecosystem.

Also, although the concept of a quality objective has a similar meaning for several international organisations, the procedures proposed for the setting of quality objectives can be extremely different. In some cases these procedures are supported by relatively rigid rules such as the USEPA approach (Stephan *et al.*, 1985), which requires 16 acute toxicity tests, 3 chronic tests, 2 plant tests, and 1 bioaccumulation study. The procedures proposed by the OECD and practised, for example, in the Netherlands for assessment of the effects of priority chemicals are very similar in design as well as in their quantifiable details.

An interesting approach to drawing up quality criteria to protect the ecosystem is that proposed by Van Straalen and Denneman (1989). The method is based on two main assumptions:

> NOELs for single species are naturally independent and represent estimates of sensitivity;
> the ecosystem is protected if NOEL is not exceeded for a large percentage of species (arbitarily 95%).

A substantial improvement in this method, as compared with the EPA approach, is the application of a sound statistical method for the calculation of the hazardous concentration for sensitive species (HCS); that is, the concentration at which the NOEL for the most sensitive species can be exceeded with a specific probability (e.g. 5%).

A conceptual objection to this approach is the arbitrariness of the assumption that 95% of species protected are 'safe' for the ecosystem. For example, the CSTEE/EEC and the EIFAC/FAO definitions of quality criteria or objectives implicitly refuse to accept the concept of the loss of part of the ecosystem. Moreover, a practical objection is the need for NOELs for data derived from a relatively high number of species (not fewer than seven) to achieve a reliable estimate of the HCS. This implies a huge amount of experimental work, unpracticable for a large number of compounds. A more realistic and pragmatic approach is that followed by the CSTEE/EEC Ecotoxicity Section (1994), based on the use of application factors in combination with a critical review of available experimental data.

In view of the variability of toxicity test results inherent in the multitude of experimental detail, the Committee decided to use application factors that express orders of magnitude of extrapolation – namely 1000, 100, and 10. In its present standard procedure of establishing WQOs, due consideration is therefore given to the best available toxicity data by applying the factors as follows (CSTEE/EEC, 1994):

> 1000 to the lower end of the acute $L(E)C_{50}$ range, when the data available are few, or the range of organisms is narrow, bearing in mind that outlier values may be due to error or experimental conditions that deviate too much from real world conditions;
>
> 100 to the lower end of the range of acute $L(E)C_{50}$s when there is an extensive database covering a (phylogenetically) wide range of test species, or to the lower end of the chronic $L(E)C_{50}$s, or NOEL values when few data are available;
>
> 10 to the lower end of (apparent) chronic NOEL data determined by a sufficient and representative number of tests.

Taking into account the experimental uncertainties and the variability mentioned above, the extrapolated figure is subsequently rounded to the nearest order of magnitude.

In using these extrapolation rules, it was clear to the Committee that the establishment of WQO is not merely a mathematical exercise, but always remains a case-by-case consideration, which calls for expert judgement.

The setting of acceptable daily intake and maximum allowable concentration

In order to protect human beings from exposure to toxic chemicals, in particular via food or drinking water, the FAO/WHO Joint Expert Committee on Food Additives developed the concept of an acceptable daily intake (ADI) (FAO/WHO, 1968). An ADI is defined as the daily intake of a chemical that, during an entire lifetime, appears to be without appreciable risk on the basis of all the known facts at the time. It is expressed in milligrams of the chemical per kilogram of body weight (mg/kg) per day. The procedure for setting an ADI is based on the availability of a NOEL determined through an adequate set of experimental data. The NOEL is then divided by a series of safety factors with respect to:

extrapolation from animals to humans;

extrapolation from large experimental doses to the small doses of real exposure;

extrapolation from a small number to a large number of organisms;

nature of toxicity and mode of action;

uncertainty due to the quantity of data used to determine the NOEL.

Starting from ADIs, maximum allowable concentrations (MACs) for food and water can be calculated.

To do that, some assumptions are needed.

(1) The characteristics of an 'average adult man' are defined as follows:

body weight: 70 kg

respiration: 20 m^3/day

drinking-water consumption: 2 l/day

Moreover one could hypothesise a total consumption of food of animal origin of about 250 g/day and of vegetal origin of about 1 kg/day.

(2) The main exposure routes of toxic chemicals to human beings are via air, water and food. The role of each different exposure route in determining the total daily intake (TDI) is very variable with respect to the properties and the environmental behaviour of the various chemicals. For highly volatile compounds, the air contribution is relatively high; for highly soluble chemicals water is more important. For example, the relative roles of the three exposure components estimated for polyaromatic hydrocarbons (PAHs) – typical low solubility, low volatility, high bioaccumulation compounds – is the following (WHO, 1984):

water: 0.1% of the TDI
air: 0.9% of the TDI
food: 99% of the TDI

For pesticides the contribution of water to the TDI has been assumed to be between 1% (for high lipophilic compounds: log $K_{ow} > 3.5$, see p. 88) and 10% (for low lipophilic compounds: log $K_{ow} < 3.5$). For most herbicides a value of 10% is suitable (WHO, 1984, 1987).

On these bases, MACs for drinking water can be calculated as follows:

$$MAC = \frac{ADI \times 70 \text{ kg}}{2 \, l/\text{day}} \times A$$

where A is the contribution of water to the TDI (typically 10% for herbicides).

The assessment of exposure

Factors regulating the biogeochemical cycle of xenobiotics

The distribution and fate of a chemical substance discharged into an environmental compartment (air, water, soil) follows a specific biogeochemical cycle dependent not only on diffusion and transport patterns within the single compartment, but also on partition processes among the various compartments. Moreover, within each compartment the chemical is subject to transformation processes (degradation, chemical reactions). A scheme of the main mechanisms regulating distribution of a chemical in the environment is shown in Figure 4.2.

The biogeochemical cycle of a chemical is a function of environmental parameters and of the properties of the substance itself. In particular, distribution among the various compartments is regulated by the physico-chemical properties of the chemical substance, representing essentially partition coefficients among the different phases. It follows that a few molecular parameters regulate environmental partitioning of organic substances and, as a preliminary step, a rough evaluation of distribution can be derived from single molecular properties. For example, a highly soluble substance will obviously partition mainly into water, whereas a highly volatile one will go principally into the air. In addition, many environmental factors (temperature, moisture, etc.) will affect the molecular properties (solubility, degradability, etc.).

To summarise, the environmental fate of chemical substances is the result

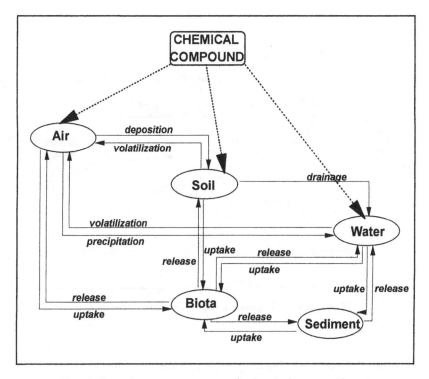

Fig. 4.2. The main transport patterns of a chemical compound moving among environmental compartments.

of a sum of partition, mass transport and degradation processes dependent on environmental and molecular factors related directly to one another.

The behaviour of a chemical in soil

Soil is a complex matrix consisting of air pockets, water, mineral matter and organic matter. It can vary enormously in its composition and texture and consists of various layers with different properties. The upper layer, which is more directly involved in the input of chemical substances and where distribution and fate processes are more complex, is the so-called vadose or unsaturated zone. This means that the pore spaces in the soil materials are not fully filled with water. Below the vadose zone, there is the saturated zone, where pore spaces are completely filled with water. The top of the saturated zone is the water table, corresponding to the level to which water will rise at atmospheric pressure in a hole dug in the earth.

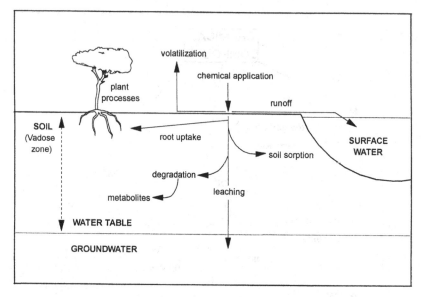

Fig. 4.3. A scheme of the major transport and fate patterns of a chemical substance in the soil environment.

The term 'groundwater' usually refers to the water below the water table. Therefore the transport of chemical substances into groundwater depends essentially on mechanisms occurring in the vadose zone.

Fate and transport of chemicals in the soil environment are controlled by complex processes and dynamic interactions (Figure 4.3). In accordance with the scheme of Donigian and Rao (1987), the soil processes can be divided into the following categories:

(1) Transport by volatilization to the atmosphere, by runoff and erosion to surface waters, by leaching to groundwater.

(2) Sorption and partition, depending on the physico-chemical properties of the substance, regulating the affinity for the organic or mineral component of the soil.

(3) Transformation/degradation processes: biodegradation, chemical hydrolysis, oxidation–reduction reactions and photolysis, the last only at the surface of the soil. Biological transformations comprise the main degradation pathway in the soil layer, where there is an active bacterial community, possibly up to some tens of centimetres deep.

(4) Volatilization defined as the loss of a chemical in vapour form from soil or plant surfaces; this depends on the physico-chemical properties of the substance, on environmental parameters (temperature, wind, etc.) and on use and application patterns (sprayed, applied on the soil surface, etc.).

(5) Plant processes. In plants chemical substances enter, are transported, are absorbed into organic matter and bioaccumulated, are metabolised and degraded. In agricultural areas, plant processes may therefore play a significant role in the fate of pesticides.

The ultimate distribution of chemicals in the soil is determined by dynamic, temporal, and spatial interactions among these processes.

The prediction of exposure

Predictive approaches for the evaluation of environmental exposure to potentially dangerous chemicals are essential for a number of reasons. From a practical point of view, environmental monitoring can be performed only *a posteriori*, after the emission of contaminants and, in extreme cases, after the occurrence of environmental damage. Therefore monitoring could allow planning of recovery measures but not any type of prevention.

Scientifically, the measurement of a given environmental concentration gives part of the picture of a punctual situation, in time and space, but does not give information about the environmental processes producing it. Knowledge of the main features of the biogeochemical cycle allows the development of conceptual instruments able to describe and predict distribution and fate patterns; from this, the presence of a chemical in the ecosystem and its movement over time can be reconstructed. Moreover, predictive approaches represent the only possibility for planning suitably preventive measures against any foreseeable risk. In the last few years, several predictive approaches have been developed at different levels of complexity and descriptive/predictive precision.

The role of molecular properties

A first, very simple approach, rough but practical at least for preliminary screening evaluations, can be derived from the quantification of the main molecular characteristics regulating environmental partitioning of a chemical. The characteristics needed and their environmental implications are as follows:

Water solubility (S) quantifies the affinity of a substance for the aqueous or water compartment. Solubilities for the more common pesticides range from several grams per litre to levels as low as 10^{-6} g/l (organochlorines, pyrethroid insecticides).

Vapour pressure (VP) indicates volatility and, therefore, the affinity for the air, a parameter actually better quantified by the Henry's Law constant. The most commonly used unit to measure the vapour pressure of pesticides is the Pascal (1 Pa = 0.0075 mmHg). With the exception of a few compounds used as fumigants, pesticides generally show relatively low volatility and the range of vapour pressure is roughly between 10 and 10^{-9} Pa.

Henry's Law constant (H) could be expressed as the ratio between vapour pressure and water solubility ($H = VP/S =$ Pa m³/mol). In practice, H is proportional to the partition coefficient between air and water. Therefore H can be assumed to be an index of the affinity for air. Henry's constant usually ranges between 10^5 and 10^{-9} Pa m³/mol, but pesticides show values above 10 in only a few cases.

Octanol/water partition coefficient (K_{ow}) quantifies the lipophilicity of a substance and is therefore assumed to be an index of the ability to pass through biological membranes and to bioaccumulate in living organisms. It is therefore used as a measure of the affinity for the biota. For most pesticides K_{ow} ranges from 10^6 to 10^{-2}. The following is a commonly used, unofficial classification scheme:
bioaccumulating substances $\log K_{ow} > 3.5$
low bioaccumulation potential $3 < K_{ow} < 3.5$
non-bioaccumulating substances $\log K_{ow} < 3$

Octanol/air partition coefficient (K_{oa}) has been defined recently as the key parameter for the evaluation of bioaccumulation in plants (Paterson *et al.*, 1991). The octanol/air partition coefficient can be obtained from:

$$K_{oa} = (K_{ow}/H) RT$$

where R is the gas constant (R = 8.314 Pa m³/mol per K) and T is the absolute temperature (T in K = 273.15 + t°C). Usually K_{oa} for pesticides ranges from 10^2 to 10^{10}.

Organic carbon sorption coefficient (K_{oc}) is usually assumed to be an index of soil affinity. In practice it represents the sorption coefficient for the organic carbon of the soil and it is directly related to K_{ow}. It can be experimentally determined or estimated from K_{ow}. The most commonly used equation for the calculation of K_{oc} is (Karickhoff, 1981):

$$\log K_{oc} = \log K_{ow} - 0.21$$

This definition of K_{oc} indicates a coefficient dependent only on the partition properties of a substance between two phases (organic carbon and water) and independent of other possible processes of sorption or other interactions with the inorganic matrix of the soil.

The true partition coefficient between soil and water ($K_p = C_s/C_w$) where C_s and C_w are the concentrations in soil and water, respectively) depends on the amount of organic carbon in the soil. Thus K_p can be calculated as:

$$K_p = K_{oc} \times F_{oc}$$

where F_{oc} is the fraction of organic carbon in soil.

The use of K_{oc} as an index of affinity for soil is suitable for relatively hydrophobic and non-ionic substances. For these compounds interactions with the inorganic matrix of the soil are negligible and soil sorption can be assumed to be determined only by partition within the organic matrix. For highly polar or ionised chemicals, electronic interactions with the inorganic matrix must be taken into account and K_{oc} is not suitable for predicting soil sorption. This is the case for a few pesticides, such as the herbicide paraquat, which is strongly bound to soil as a result of its cationic properties, even if it shows high water solubility and low K_{oc}.

A tentative qualitative scheme for the classification of the affinity of chemicals for the different environmental compartments is shown in Table 4.3. The classification is more reliable for the extreme values of the

Table 4.3. *Affinity of an organic chemical for the different environmental compartments according to the main molecular parameters (from Vighi and Di Guardo, 1995)*

Affinity for the compartments	Water S (g/l)	Air H (Pa m^3/mol)	Soil $\log K_{oc}$	Animal biomass $\log K_{ow}$	Vegetal biomass $\log K_{oa}$
Very high	>1	>10	>5	>5	>8
High[a]	1–10^{-2}	10–10^{-1}	5 4	5 3.5	8–7
Average[a]	10^{-2}–10^{-3}	10^{-1}–10^{-2}	4–2	3.5–3	7–5
Low[a]	10^{-3}–10^{-5}	10^{-2}–10^{-4}	2–1	3–1	5–4
Very low	$<10^{-5}$	$<10^{-4}$	<1	<1	<4

Note:

[a] Influenced by other parameter values.

parameters because of negligible interactions among other properties when one is strongly prevalent (Table 4.3). A solubility higher than 1 g/l indicates clearly that this property is the main driving force and the affinity for water is always very high. On the other hand, $\log K_{ow}$ and $\log K_{oc}$ values around 6 unequivocally indicate very high bioaccumulation and soil sorption potential. At intermediate levels the simultaneous effect of various properties is more complex and the classification must be taken as purely indicative.

Leaching indexes and ranking systems

A further step, somewhat more advanced than the simple evaluation of single molecular parameters, is represented by comparative indexes and ranking systems. This approach requires a few input data, either molecular properties or environmental parameters, and is based on simple algorithms that cannot be assumed to be true models. These indexes produce non-quantitative values that allow the comparison of several compounds and the hazard ranking for one or more environmental compartments. In particular, most of these systems were produced specifically for groundwater and therefore allow a classification of the leaching capability of chemical substances. Vighi and Di Guardo (1995) have given a critical review of the most common currently used leaching indexes.

An example is the GUS (groundwater ubiquity score) index, based on K_{oc} and half-life in soil $(t_{1/2})$ (Gustafson, 1989):

$$GUS = \log t_{1/2}(4 - \log K_{oc})$$

On the basis of the previous algorithm, pesticides can be classified as follows:

non-leachers: $GUS < 1.8$

transition compounds: $1.8 \leq GUS \leq 2.8$

leachers: $GUS > 2.8$

The reliability of the extremely simple algorithm of the GUS index was successfully validated by applying it to a complex model called GLEAMS (Goss, 1992).

The GUS index, like most leaching indexes, does not take into account pesticide application rate. Therefore these kinds of index measure an intrinsic leaching capability of the chemicals instead of a realistic pollution potential of the applied compound. This is a severe limitation to the practical implications of these indexes, particularly as regards the big differences in application rates of many new-generation active ingredients (e.g. the sulfonylurea herbicides), which are applied at the level of grams per hectare, whereas traditional compounds are applied at the level of kilograms per hectare (e.g. triazine herbicides). A possible improvement in the practical reliability of the index could be the introduction of the application rate into the algorithm, as in the modified GUS index (GUS_m):

$$GUS_m = \log t_{1/2}(4 - \log K_{oc})A_r$$

where A_r is the application rate, expressed in kilograms of active ingredient per hectare.

Partition analysis and multicompartmental models

The concept of evaluative models for the prediction of partition among environmental compartments was introduced by Baughman and Lassiter (1978). This kind of model considers the comprehensive role of the chief molecular properties and their interactions, giving a complete picture of the environmental distribution of a chemical substance rather than the two-phase partition described by simple partition coefficients. These models were originally developed as 'instruments for thinking' and for the indication of general trends in environmental partitioning, not as quantitative tools for the prediction of environmental concentrations. Nevertheless, they proved to be highly versatile instruments and in the last few years several 'site-specific' models have been developed.

At present multicompartmental partition analysis models are considered to be the most effective tools for the quantitative prediction of environmental distribution and the fate of chemical substances. Several models have been developed at different levels of complexity and validated at different spatial scales (Vighi and Di Guardo, 1995). Relatively complex models, such as the pesticide root zone model or PRZM (Carsel *et al.*, 1985), which require a detailed description of environmental features as input data, can give very good predictions at the field scale (not more than a few hectares) (Carsel *et al.*, 1986), but they fail on a larger scale. There are no theoretical objections to the application of such models on a wider area, but the variability of environmental parameters makes it practically impossible to obtain a description which is as detailed and accurate as the model requires.

More simple models, requiring only an approximate description of the main driving forces as input data, produce less precise results but their versatility allows their application to relatively non-homogeneous areas and, therefore, on a larger scale. Simple runoff models derived from the original 'fugacity approach', were developed at the University of Toronto by Mackay and co-workers, but too technical to be described here. (Mackay, 1991), have been successfully validated. They indicate an acceptable predictive capability (within one order of magnitude) in several experimental areas of hundreds of hectares (Di Guardo *et al.*, 1994), and even as large as the drainage area of a small river, with a surface of more than $100 \, km^2$ (Barra *et al.*, 1995).

The problem of persistence

Information about persistence is essential for the environment risk assessment of chemical substances. Persistence is needed as input for all predictive approaches, from simple leaching indexes to more complex models. Nevertheless, the availability of reliable persistence data is, at present, the weakest link in the prediction of the environmental fate of chemicals.

This can be explained in part using the relative complexity of degradation

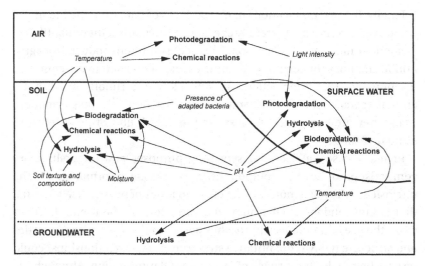

Fig. 4.4. A scheme of the major degradation processes occurring in the main environmental compartments and of the influence of environmental variables.

patterns (Figure 4.4), determined by a number of processes (biodegradation, hydrolysis, photodegradation, chemical oxidation, etc.) with variable roles in different environmental compartments. Moreover each process is a function of the intrinsic stability of the molecule and of a number of environmental variables (temperature, moisture, pH, soil texture and composition, etc.).

In the upper layers of the soil, biodegradation is the main degradative process for most compounds and is due to the presence of an adapted bacterial population. Thus, a pesticide can be degraded very rapidly in soil treated for a long time with the same compound, but very slowly when the soil is treated for the first time. Moreover, in the different soil layers, different degradative processes occur and the half-life varies widely. For example, the half-life of atrazine resulting mainly from biodegradation in the surface soil layers is around some tens of days, whereas in deep soil, and in groundwater only hydrolysis and other abiotic processes can occur and the half-life is of the order of years.

Precise persistence values can therefore be obtained experimentally only case by case and average values of soil $t_{1/2}$ are merely rough indications. In many cases it is better to indicate a range in which the most probable value could fall.

Finally, the real implications and usefulness of data from the literature, when available, must be carefully checked, particularly if they quantify the residence time of pesticides in soil. These data are often produced for agronomic and not ecotoxicological purposes and are related to the complete disappearance of the molecule as a result of leaching, runoff, volatilization and other transport patterns, and not only reaction and transformation. Thus, they are completely useless for the description of persistence in evaluative models.

In many cases, at least for screening purposes and for preliminary comparisons of several compounds, approximate information on the intrinsic stability of a molecule, taken as an index of persistence potential that is independent of environmental variables, can be useful. In these cases the use of predictive approaches based on the molecular properties and structure (QSAR; quantitative structure–activity relationships) could be very helpful in the absence of experimental information. Although the application of QSARs for the prediction of persistence has not yet been developed for screening as it has for other ecotoxicological aspects (e.g. prediction of toxic effects or bioaccumulation), in the last few years there has been some promising progress (Tremolada *et al.*, 1991; Vasseur *et al.*, 1993; Macalady and Schwarzenbach, 1993).

How to perform a risk assessment

In conclusion, the risk assessment for a potentially dangerous chemical introduced into the environment must be based on information related to the substance, to the environmental characteristics, and to the population potentially exposed (Figure 4.5). The effects of the chemical on living organisms are studied by means of toxicological testing, bioaccumulation, etc. and a quality objective for the environment or for a natural resource (e.g. drinking water) is set in order to protect the population (human beings or wildlife) that may be exposed.

On the quantitative side, an environmental concentration is either measured by experimental monitoring or evaluated by means of predictive models. For both, experimental or predictive, approaches information on the use patterns of the substance, on its intrinsic properties, on its persistence in relation to the different environmental compartments and on the main environmental processes influencing distribution and fate of the chemical are needed. In the case of predictive models, this information

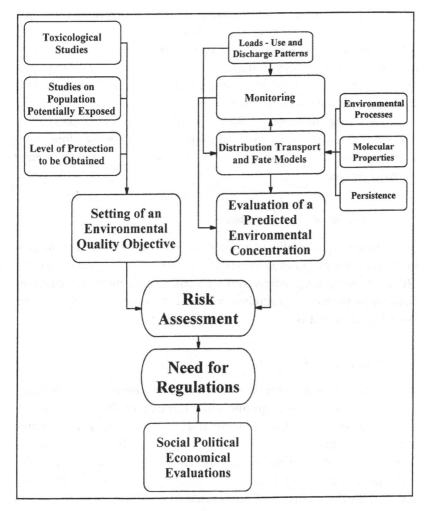

Fig. 4.5. A schematic procedure for environmental risk assessment.

represents the essential set of input data, more or less detailed and precise depending on the approach used. In the case of experimental measurement, it is also useful to plan appropriate monitoring activity.

As previously mentioned, predictive approaches are the obvious tool for preventive evaluations. For chemicals already present in the environment, distribution and fate models are extremely useful, possibly in combination with experimental measurements. Monitoring work should be planned on the basis of the results of environmental models and experimentally measured values could be used to confirm and validate the reliability of the predictive approaches.

Risk assessment will result from a comparison between the environmental quality objectives or standards and the actual (measured or predicted) environmental concentration and this comparison will indicate the need for possible control measures.

Finally, taking into account political, social and economic aspects, regulatory actions would be defined. This outline procedure therefore represents the contribution of ecotoxicological science to environmental decision-making.

Risk assessment of atrazine

An ecotoxicological profile of atrazine and some possible substitutes has been reported by Vighi and Zanin (1994). On the basis of the general principles described in the previous sections, a risk assessment for atrazine, particularly in respect of groundwater intended for human consumption, could be performed as follows.

Exposure assessment

This chapter does not aim to give a quantitative evaluation of atrazine concentration in Italian groundwater through predictive or experimental approaches. A qualitative assessment of atrazine leaching potential can, however, be made and compared with other compounds used for weed control on maize. For screening purposes the groundwater pollution potential can be calculated by means of the GUS index, modified by the inclusion of the application rate. In Table 4.4 standard and modified GUS indexes are reported for atrazine and for the most important herbicides used on maize in Italy after the atrazine ban (Sbriscia Fioretti *et al.*, Chapter 2, this volume). The modified GUS index indicates a high potential of groundwater pollution for atrazine and for some substitute compounds such as alachor, metolachlor, butylate, dicamba and EPTC.

It is interesting to note that the sulphonylureas (primisulfuron and rimsulfuron) have a very high leaching capability, indicated by the high value of the unmodified GUS index, but, owing to the extremely low application rate, the groundwater pollution potential is negligible. Before the ban,

Table 4.4. *Values of the GUS index, of the usual application rate for the active ingredient and of the modified GUS index for atrazine and for its major substitutes*

Compounds	GUS	Application rate (kg a.i./ha)	GUS_m
Atrazine	2.9	1.5	4.35
Alachlor	1.8	2	3.6
Butylate	1.9	5	9.5
Dicamba	5.7	0.2	1.14
EPTC	1.9	5	9.5
Linuron	3.6	0.5	1.8
Metolachlor	2.3	2	4.6
Pendimethalin	−1.3	0.6	−0.78
Primisulfuron	6.65	0.016	0.1064
Rimsulfuron	6.16	0.015	0.0924
Terbutylazine	3.1	0.6	1.86

Note:
a.i., active ingredient.

atrazine was used on 80% to 90% of the Italian maize crop area, so it is not surprising that groundwater pollution is widespread. Atrazine was frequently found in groundwater resources used for drinking-water supplies but the levels measured were usually relatively low. The majority of positive analytical data available in the three major regions of maize production was in the range 0.1 to 1 μg/l (Vighi and Zanin, 1994).

During the period of regional bans (1987–1990), concentrations above 1 μg/l were determined in some wells and this led to the ban being imposed throughout the whole municipal territory. Nevertheless values above 2 μg/l were detected only sporadically.

Toxicological assessment

Atrazine is slightly toxic to mammals. The NOELs figures obtained during two years of feed trials range from about 1 to about 4 mg/kg daily for rats and dogs respectively (Tomlin, 1994). It can act as a weak, non-genotoxic carcinogen in rat, with NOEL of 0.7 mg/kg. The ADI for humans is 0.7 μg/kg body weight and the drinking-water guideline proposed by the World Health Organization is 2 μg/l (WHO, 1987). On the other hand, the acceptable limit for pesticides in drinking water, proposed by Directive 80/778/EEC, is as low as 0.1 μg/l, independent of their toxicological properties. The implications and validity of this assessment are discussed in the preface (Vighi *et al.*, this volume).

In conclusion, atrazine has been found to be a regular contaminant of groundwater. The detection of concentrations above the EEC standard was very frequent in northern Italy before the ban and due to its persistence in deep soil and groundwater (half-life in the order of years), such detections are even now not rare events. This justifies the need for control measures at least in areas at higher risk. The total national ban was the consequence of an emotional situation, determined more by public fear than by sound scientific evaluation. Therefore, probably, it was not the best solution.

In fact, if the actual concentrations in groundwater are compared with the WHO Guidelines, conditions of real risk for human beings were sporadic and limited to restricted areas. A comprehensive strategy of weed control management on maize, based on sound agronomic, environmental and economic grounds, could result in better protection of the groundwater from herbicide contamination and could reasonably meet the standard for drinking water in the medium term.

References

Barra, R., Vighi, M. and Di Guardo, A. (1995). Prediction of surface water input of chloridazon and chlorpyrifos in an agricultural watershed in Chile. *Chemosphere*, **30**, 485–500.

Baughman, G. L. and Lassiter, R. R. (1978). Prediction of environmental pollutant concentration. In *Estimating the Hazard of Chemical Substances to Aquatic Life*, ed. J. Cairns Jr, K. L. Dickinson and A. W. Maki, pp. 000–00. ASTM STP 657, Philadelphia.

Blank, H., Dave, G. and Gustafsson, K. (1978). *An Annotated Literature Survey of Methods for Determination of Effects and Fate of Pollutants in Aquatic Environments*. Swedish Board for Technical Development, Solna, Sweden.

Carsel, R. F., Mulkey, L. A., Lorber, M. N. and Baskin, L. B. (1985). The pesticide root zone model (PRZM): a procedure for evaluating pesticide leaching threats to ground water. *Ecological Modelling*, **30**, 46–49.

Carsel, R. F., Nixon, W. B. and Ballantine, L. B. (1986). Comparison of pesticide root zone model predictions with observed concentrations for the tobacco pesticide metalaxyl in unsaturated zone soils, *Environmental Toxicology and Chemistry*, **5**, 345–53.

Casarett, L. J. and Doull, J. (1975). *Toxicology. The Basic Science of Poisons*. Macmillan Publ. Co., Inc., New York.

CSTEE/EEC (1987). Internal report CSTE/877101/XI from Directorate/EEC General for Environment, Nuclear Safety & Civil Protection, DG XI/A/2, Brussels.

CSTEE/EEC Ecotoxicity Section (1994). EEC Water Quality Objectives for chemicals dangerous to aquatic environments. *Reviews on Environmental Contamination and Toxicology*, **137**, 83–110.

Di Guardo, A., Calamari, D., Zanin, G., Consalter, A. and Mackay, D. (1994). A fugacity model of pesticide runoff to surface water: development and validation. *Chemosphere*, **28**, 511–32.

Donigan, A. S. and Rao, P. S. C. (1987). Overview of terrestrial processes and modeling. In *Vadose Zone Modeling of Organic Pollutants*, ed. S. C. Hern and S. M. Melancon, pp. 3–36. Lewis Publishers, Inc., Chelsea, MI.

EEC (1973). Declaration from the EEC Council of Ministers of 22 November 1973 concerning the first Environmental Action Programme for the European Communities, *Official Journal of the European Community* C 112/1.

FAO/WHO (Food and Agriculture Organization/World Health Organization) (1968). *Specifications for the Identity and Purity of Food Additives and their Toxicological Evaluation*. Joint FAO/WHO Expert Committee on Food Additives. FAO Nutrition Meetings, Report series no. 44.

Goss, D. W. (1992). Screening procedure for soils and pesticides relative to potential water quality impacts. *Weed Technology*, **6**, 701–8.

Gustafson, D. I. (1989). Ground water ubiquity score: a simple method for assessing pesticide leachability. *Environmental Toxicology and Chemistry*, **8**, 339–57.

Karikhoff, S. W. (1981). Semiempirical estimation of sorption of hydrophobic pollutants on natural sediments and soils. *Chemosphere*, **10**, 833–49.

Lloyd, R. (1991). Some ecotoxicological problems associated with the regulation of PMPs. In *Persistent Pollutants: Economy and Policy*, ed. H. Opshoor and D. Pearce, pp. 45–62. Kluwer Academic Publishers, Dordrecht.

Lloyd, R. and Calamari, D. (1987). Water quality criteria for European freshwater fish: the role of EIFAC (FAO). *Water Quality Bulletin*, **12**, 100–3.

Macalady, D. L. and Schwarzenbach, R. (1993). Predictions of chemical transformation rates of organic pollutants in aquatic systems. In *Chemical Exposure Prediction*, ed. D. Calamari, pp. 27–46. Lewis Publishers, Inc., Boca Raton, FL.

Mackay, D. (1991). *Multimedia Environmental Models, The Fugacity Approach*. Lewis Publishers, Inc., Chelsea, MI.

OECD (Organization for Economic Co-operation and Development) (1992). *Environment Monograph* no. 59, from the Arlington Workshop, 10–12 December 1990. Report no. OCDE/GD(92)169 from the Environment Directorate, Paris.

Paterson, S., Mackay, D., Bacci, E. and Calamari, D. (1991). Correlation of the equilibrium and kinetics of leaf-air exchange of hydrophobic organic chemicals. *Environmental Science and Technology*, **25**, 866–71.

Slooff, W., Van Oers, J. A. and De Zwart, D. (1986). Margins of uncertainty in ecotoxicological hazard assessment. *Environmental Toxicology and Chemistry*, **5**, 841–2.

Stephan, C. E., Mount, D. I., Hansen, D. J., Gentile, J. H., Chapman, G. A. and Brungs, W. A. (1985). *Guidelines for Deriving Numerical National Water Quality Criteria for the Protection of Aquatic Organisms and their Uses*, US Environmental Protection Agency, PB85-227049, Springfield, VA.

Tomlin, C. (ed.) (1994). *The Pesticide Manual*. British Crop Protection Agency, Farnham.

Tremolada, P., Di Guardo, A., Calamari, D., Davoli, E. and Fanelli, R. (1991). Mass-spectrometry-derived data as possible predictive method for environmental persistence of organic molecules. *Chemosphere*, **24**, 1473–92.

USEPA (United States Environmental Protection Agency) (1987). *US Federal Register*, **50**, 30784.

Van Straalen, N. M. and Denneman, C. A. J. (1989). Ecotoxicological evaluation of soil quality criteria, *Ecotoxicology and Environmental Safety*, **18**, 241–51.

Vasseur, P., Kuenemann, P. and Devillers, J. (1993). Quantitative structure biodegradability relationships for predicting purpose. In *Chemical Exposure Prediction*, ed. D. Calamari, pp. 47–62. Lewis Publishers, Inc., Boca Raton, FL.

Vighi, M. and Di Guardo, A. (1995). Environmental distribution and fate and exposure prediction. In *Pesticide Risk in Groundwater*, ed. L. Bergman and M. Pugh, pp. 73–100. Lewis Publishers, Inc., Boca Raton, FL.

Vighi, M. and Zanin, G. (1994). Agronomic and ecotoxicological aspects of herbicide contamination of groundwater in Italy. In *Environmental Toxicology, Economics and Institutions – The Atrazine Case Study*, ed. L. Bergman and M. Pugh, pp. 111–40. Kluwer Academic Publishers, Dordrecht.

WHO (World Health Organization) (1984). *Guidelines for Drinking-Water Quality*, vol. 1 *Recommendations*. WHO, Geneva.

WHO (World Health Organization) (1987). *Drinking-Water Quality: Guidelines for Selected Herbicides*. Environmental Health Series, 27, WHO, Regional Office for Europe, Copenhagen.

5   The value of changes in health risks: a review

Magnus Johannesson and Per-Olov Johansson

Introduction

There is a large and fast-growing literature on the economic value of changes in health risks. The theoretical foundations for willingness-to-pay (WTP) measures of risk changes have been explored and seem to be quite solid (see, for example, Jones-Lee, 1976; Rosen, 1988; Viscusi, 1992, 1993; Johansson, 1995). There are also quite a few empirical studies of the value of changes in health risks. In fact, several different methods, based on behaviour either in actual markets or in hypothetical or constructed markets, have been used to estimate the value of what is known as a 'statistical life'.

This chapter is structured as follows. In the first section we introduce a simple model in order to provide definitions of the value of changes in health risks and other concepts which are used in later sections. We then go on to present the empirical methods which have been used to estimate the value of risk changes. Available empirical results are summarised, and we also provide a brief comparison of the value of a statistical life according to different methods and studies. The chapter ends with a few remarks on the evaluation of changes in groundwater quality.

Some basic definitions

In order to define WTP measures for changes in health risks, we will use a very simple model. Let us consider an individual who consumes a single private commodity and faces an exogenously fixed survival probability. His indirect utility function is written as follows:

$$V = V(\pi, y) \tag{5.1}$$

where V is the level of utility, $V(.)$ is the indirect utility function, π is the survival probability and y is income.

Assume that a project causes a small change in the individual's survival probability π. We want to find the individual's monetary valuation of the change in π. Totally differentiating equation (5.1), adjusting income so as to keep the individual's level of utility unchanged, yields after straightforward calculations:

$$-dy/d\pi = V_\pi/V_y = \text{MRS}_{y_\pi} \tag{5.2}$$

where $V_\pi - \delta V(.)/\delta\pi$, $V_y = \delta V(.)/\delta y$, and MRS_{y_π} is the marginal rate of substitution between income and health risk. Equation (5.2) highlights the fact that there is normally a trade-off between income and risk. That is, an individual is prepared to accept a slightly higher death risk (lower survival probability) in exchange for a higher income and vice versa.

Rearranging equation (5.2), one obtains the individual's willingness to pay for a small change in the survival probability:

$$dy = d\text{CV} = d\text{EV} = - [V_\pi/V_y]d\pi = - \text{MRS}_{y_\pi}d\pi \tag{5.3}$$

where dCV denotes the (marginal) compensating variation and dEV is the (marginal) equivalent variation. Equation (5.3) expresses the (marginal) willingness-to-pay for a small change in the survival probability π. If the individual is risk-averse with respect to health risks ($V_{\pi\pi} < 0$), his[2] WTP for a ceteris paribus reduction in the death risk will increase with the level of death risk (i.e. $1 - \pi$). That is, the higher is the initial risk level, the more the individual is willing to pay for a small risk reduction.

In many cases, an individual is able to affect the probability of experiencing a particular event. For example, he can buy fire alarms in order to reduce the risk of a fatal fire or shift from 'risky' but cheap tap water to less risky but more expensive mineral water. Let us therefore augment our simple model so as to become able to handle endogenous risks. Assume that the individual consumes a single private good x and purchases a health good X in order to increase his survival probability. His direct utility function, $U(.)$ is assumed to be of the form $U[\pi(X), x]$. Thus, there is a production function $\pi(X)$ 'translating' the health good X into a survival probability, i.e. $\pi = \pi(X)$. As is further explained in the Appendix to this chapter, the indirect utility function generated by this model is as follows:

$$V(p, P, y) = U[\pi(p, P, y), x(p, P, y)] \tag{5.4}$$

where π is the survival probability, p is the price of the consumption good x, P is the price of the health good X and y is exogenous income. Note that we can alternatively interpret x, p, X and P as vectors, although for simplicity we stick here to the single goods interpretation.

The individual's willingness to pay for a marginal increase in his survival probability is easily shown to be equal to the marginal cost of achieving the increase, assuming here an interior solution. This is so because the individual has already chosen the optimal level of π subject to his budget constraint, implying that the marginal gain from a small increase in π is equal to the marginal cost of achieving the increase in π:

$$[V_\pi(.) / V_y(.)] \delta\pi(.) / \delta X = P \qquad (5.5)$$

where $V_x(.)$ is the marginal utility of risk reductions, i.e. $\delta U(.) / \delta\pi$. Thus, the WTP for the risk reduction ($d\pi = (\delta\pi(.) / \delta X) dX$) caused by a small increase in X is equal to its market price P (times the change in X, i.e. dX). This can most easily be seen by comparing equations (5.3) and (5.5) noting that $d\pi = (\delta\pi(.) / \delta X) dX$. Market data can therefore sometimes be used to estimate the WTP for small risk reductions.

Some measures, for example installing a fire alarm in one's house, will affect the survival probabilities of all household members. Replace therefore the individual utility function by a well-behaved household welfare function:

$$U = [\pi(X), x] \qquad (5.6)$$

where $U(.)$ is the direct utility function, $\pi(X) = [\pi_1(X), \ldots, \pi_m(X)]$ is a vector of the survival probabilities of the m different household members, X is a public health good at the household level and x is a (vector of) private consumption good(s). According to this formulation, the health good X is a collective or public good at the household level. Investing in X will change the survival probabilities of all household members. The household will invest in (purchase) X until the *sum* of each household member's marginal utility of a risk reduction is equal to the marginal cost of the risk reduction (in terms of private consumption of x crowded out); see the Appendix for details. The (household) indirect utility function will have prices and income as arguments, i.e. will appear identical with the one used above. This means that we can evaluate changes in P in the way proposed above, i.e. in equation (5.5).

The last case considered above illustrates *altruism* at the household level. The head(s) of the household care about the welfare of the other household members. One can also visualise cases in which people express altruism towards others than those in their own household. This may be the case for public good measures such as improved public safety. A person may be willing to contribute to such measures even if none of the members of his household will be affected. Note, however, that it is not possible to use market prices in order to evaluate safety measures (having or not having a public good property) in the presence of *inter*-household altruism. Usually, the market does not cover such motives for a willingness to pay. The reader is referred to Johannesson (1994, 1995) and Jones-Lee (1991, 1992) for further discussion of the treatment of altruism in evaluations of changes in health risks.

Many public safety and public health measures have the property that it is unknown in advance whose lives will be saved. For example, measures taken to improve the quality of a road or to increase the number of emergency vehicles in a region have this property. Suppose now that we have somehow calculated each affected individual's (marginal) non-contingent compensating variation for such a measure. There are H affected individuals and the measure saves b lives, i.e. reduces the probability of death by b/H, where b/H is taken to be a small number. Then, and as is shown in the Appendix, the *value of statistical life* is equal to $\Sigma_h dCV^h/b$, where dCV^h is the marginal compensating variation of individual h. In other words, the value of a statistical life is equal to the aggregate willingness to pay for a measure saving b lives divided by the number of lives saved (i.e. b). Alternatively, and as shown in the Appendix, the value of statistical life is given by the mean, over the H affected individuals, of their marginal rates of substitution between income and risk.

In order to provide a simple numerical illustration, suppose that a programme reduces the number of people killed in traffic accidents from 7 per 100 000 road users to 5 per 100 000 road users. Suppose that the total number of trafficants is $H = 1 000 000$ so that the programme saves 20 lives, i.e. $b = 20$. If the total willingness to pay, i.e. $\Sigma_h dCV^h$, is $30 million, then the value of statistical life is $1.5 million ($30 million/20).

We have used the marginal compensating variation to illustrate the concept of the value of statistical life. However, as is indicated by equation (5.3), in evaluating marginal changes it holds that $dCV = dEV$. Thus we can use the equivalent variation measure as well in defining the value of statistical life. This is further shown in the Appendix.

To summarise, this section has demonstrated that we can put a monetary value on the value of changes in death risks. This should come as no surprise since people often seem to act as if they are prepared to accept a higher death risk in exchange for a benefit of one kind or another. This is not to say that we have proved that a monetary valuation of changes in death risks is meaningful. But we have at least indicated sets of assumptions that make such valuations of changes in (morbidity and) mortality a consistent exercise.

Empirical methods[3]

Broadly speaking, there are two alternative approaches for empirical economic valuation of health changes and health risks; the *human captial* (HC) approach and the *willingness to pay* (WTP) approach. Both alternatives will be briefly described and compared before a closer look is taken at the estimation methods involved in the latter approach. Estimates of the value of a statistical life will also be discussed, in particular their relation to the estimation methods used.

The HC approach views the value of an individual as equal to the value of his contribution to total production and assumes that this value can be measured as his earnings. This means that the value of preventing someone's statistical death or injury is equal to the gain in the present value of his future earnings. Such a valuation has some disturbing consequences. For example, the statistical life of retired people has no value. Similarly, the implication of discounting future earnings is that the statistical life of children is likely to be of less worth than that of adults in or near their best period of earning. People whose value of production is not reflected by wage payments are also difficult to handle within the HC framework; housewives may serve as a good example. Moreover, the approach does not take into account the indirect damages due to health and injuries, e.g. pain and suffering, both those of persons directly affected and those of relatives and friends. These and other shortcomings have led to attempts to adjust estimates obtained by the HC approach, for example by valuing housewives' production by wages of domestic servants and including allowances aimed at capturing economic losses due to pain and suffering. Such attempts do not, however, solve what is usually viewed as the most important drawback of the HC approach. It is simply not consistent with the individualistic foundation of welfare economics, since it does not take people's own preferences on changes in health risks into account. Such consistency is, however, accomplished by the WTP approach.

The WTP approach assumes that a reduction in health risks is viewed by individuals in the same way as most other goods and services whose consumption affects their well-being. If so, it should be possible to observe that people make rational choices between situations involving different health risks in order to maximise their well-being subject to their budget constraints. For example, they should demand some sort of compensation in order to accept occupations involving larger health risks than other jobs, and they should choose to invest in a costly protection against a health risk only when the economic benefits of the resulting decrease in risk are at least as large as the protection costs. The existence of rational choices of these kinds is the basis for the WTP approach. In what follows we will consider the two most common methods for the estimation of the WTP for changes in health.

Indirect estimation methods

Indirect methods exploit peoples' behaviour in markets where goods related to health risks are traded. To study individuals' averting behaviour is intuitively appealing, since this is really a case where individuals buy themselves a risk reduction for money. For example, Åkerman *et al.* (1991) examined residents living in houses with indoor radioactive radiation from radon decay products. Given information on radiation levels and the health risks, the residents decide on whether they should take measures against the radiation or not. This decision obviously involves a consideration of what they are willing to pay for a radiation reduction. Using data on measures undertaken, costs, and radiation levels, Åkerman *et al.* estimated a WTP and the implied value of statistical life. Other examples of averting behaviour are purchases of smoke detectors, see Dardis (1980), and use of seatbelts, see Blomquist (1979). A number of studies of averting actions undertaken in response to groundwater contamination are summarised on pp. 113–16. It is important to note that averting behaviour implies that health risks faced by the individuals are endogenous. Crocker *et al.* (1991) analysed the consequences of endogenous risk for economic valuation, and drew attention to the fact that individuals may differ in their ability to reduce risk. This is not reflected in the notion of the value of a statistical life. It has also been argued that self-protection expenditures can be interpreted as a lower bound to the value of risk reduction. Shogren and Crocker (1991) showed that this claim is not necessarily true.

There are a number of studies using housing markets to examine health risks. These studies are based on the hypothesis that house characteristics yielding differences in health risks across houses should be reflected in property value differentials. Examples are provided by air pollution (see Portney, 1981), exposure to hazardous waste (Kohlhase, 1991), groundwater contamination (Malone and Barrows, 1990), and radon radiation (Söderqvist, 1991).

The most common indirect method, however, is to study wage differentials on the labour market. In principle, wage differentials are explained as the outcome of firms' different offers of wages depending on what health risks are involved and workers' different preferences for safety, given firms' and workers' institutional environments (see Rosen, 1986, for details). To put it simply, it is costly for firms to reduce the health risks involved in jobs, which implies a trade-off between wages and risk reduction costs; if a firm has undertaken risk reduction measures, the maximum wage it is willing to offer will be lower than otherwise. Workers differ in their view on safety, implying that their indifference curves for trade-offs between health risks and wages also differ. As a consequence one obtains a series of tangency points between the (aggregated) offer curve of firms and workers' indifference curves. The idea is to estimate this locus of tangency points, which sometimes is called the hedonic wage function. In essence, one uses regression analysis, where wages/earnings are regressed on worker and job characteristics, including risk levels. The partial derivative of the estimated hedonic wage function with respect to the risk characteristic is interpreted as the risk premium for a given level of risk (and given values of other characteristics). Given the risk premium, one can infer the value of a statistical life.

An article by Kniesner and Leeth (1991) will serve as an example of a recent wage differential study. The authors use data from the labour markets in Australia, Japan and the USA, respectively. It should be noted that they follow the normal track and use cross-industry data. An alternative, first employed by French and Kendall (1992), is to concentrate on data on a single industry. Anyhow, use of data from different countries implies that Kniesner and Leeth (1991) can examine the effects of differences in institutional environments on wage differentials. In order to illustrate the econometrics involved, let us examine their estimation of a hedonic wage function for the manufacturing industry of the USA. The authors begin by noting that an estimated hedonic wage function is sufficient for their purposes; they do not need to digress into the complicated issue of estimating underlying offer functions and indifference curves (see Epple, 1987). They then choose a semi-logarithmic form of the hedonic wage function, evidently because of its popularity in econometric studies and not its fitness for available data. The dependent variable of the hedonic wage function is defined as average weekly earnings of full-time manufacturing workers. This variable is regressed on fatality rate, injury rate, benefits from workers' compensation system measured as the maximum weekly death benefit to the surviving spouse, a dummy variable for the existence of a maximum limit of total spouse benefit, race, sex, marital status, age, education, union membership, region, the industry's new hire rate and its separation rate. This choice of variables illustrates some important issues.

First, there is always the risk that some characteristics causing differences in worker productivity are not included as independent variables (or are unobservable). Hwang *et al.* (1992) have analysed this issue within a simulation framework and concluded that

> point estimates reported in existing studies are likely to seriously underestimate the true compensating wage differentials they are intending to measure. For example, ... if (i) wages constitute 65 per cent of total compensation, (ii) differences in tastes account for 20 per cent of total wage variation, and (iii) unobserved productivity variance equals 40 per cent of total productivity variance, then workers' true valuations of life will be 10 times greater than valuations calculated from the estimated wage differentials.

The difficulty of observing all relevant data is a general problem to empirical research, but the results of Hwang *et al.* (1992) underscore the high degree of attention it deserves. Viscusi (1992) also touched upon this issue when noting that any omission of job characteristics positively correlated with job health risks would result in an overestimate of the health risk premium. Secondly, note that Kniesner and Leeth (1991) took into account the workers' compensation system. They found that this system has considerable influence on the estimates of wage differentials. In early wage differential studies the presence of compensation systems was not always considered. Thirdly, it can be questioned whether workers really have information on the health risks of the jobs for which they are applying. It has been argued that mobility among workers gives them risk experience from different firms and thus eventually they learn the true risks. Hire and separation rates may therefore be of relevance for hedonic wage functions.

Kniesner and Leeth (1991), who indeed included these rates as independent variables, noted that wage allowances for hazardous jobs may be
formalised as a result of government or union action. Such action clearly
causes wage differentials, but the problem of workers' actual perception of
job risks remains. This latter problem is related directly to the general issue
of objective versus subjective risks, an issue which is not discussed further
here.

Direct estimation methods

The direct approach to the estimation of WTP is to ask individuals how
much they are willing to pay for a risk reduction. This method to assess WTP
is usually referred to as the contingent valuation (CV) method. The first
attempt to use the CV method to value risk reductions was by Acton (1973),
who investigated the WTP for mobile coronary care units that would
decrease the risk of dying after a heart attack. Jones-Lee (1976) carried out
an early study of the value of airline safety. Both these early studies were
explorative in nature and used very small samples. Two later and much
larger studies investigating the value of risk reductions are those of Jones-
Lee *et al.* (1985) and Smith and Desvousges (1987).

Jones-Lee *et al.* (1985) used the CV method to investigate the value of
reductions in the risk of traffic deaths in the UK. A random sample of 1150
individuals was used, and 1103 interviews were carried out (i.e. a non-
response of about 5%). A follow-up study was also carried out on a sample
of 210 individuals from the original sample to test the reliability and stability of the answers. The results showed a value of the statistical life of about
£1.5 million (1982 prices) with only WTP for own risk reduction included
and about £2.0 million when the WTP for reductions in other people's risks
is included. The authors also note that the results as a whole conform to the
theoretical predictions, and that individuals seem to be able to understand
changes in probabilities.

Smith and Desvousges (1987) used the CV method to value risk changes
from hazardous waste. A representative sample of 720 households in
Boston, MA, was used in the study and 609 interviews were carried out
with these households (i.e. a non-response of about 15%). The risk of
dying from exposure was divided into two parts; the risk of exposure and
the risk of dying given that one is exposed. The sample was divided into
eight different subsamples confronted with different risk reductions. The

variation in WTP between the different subsamples was small. Contrary to the theoretical predictions the WTP for a small risk reduction was found to decrease with the level of the risk. However, within each subsample the hypothesis of equal marginal WTP over a range of risk levels could not be rejected. The regression analysis also indicated that individuals exaggerate small risks. Moreover, the WTP for a marginal risk reduction was greater than the WTP to avoid an identical marginal risk increase. The results were thus contrary to most theoretical predictions, which contrasts with the results and conclusions of the Jones-Lee et al. (1985) study. The fact that the risk was divided into two components (exposure and mortality risk if exposed) may, however, have led to problems for the individuals in interpreting the actual risk reductions. In comparing the Smith and Desvousges (1987) and the Jones-Lee et al. (1985) results it also seems as if results are consistent within samples, i.e. the WTP increases with the risk reduction within a sample, but not between samples. (The latter was not studied in the Jones-Lee et al. (1985) study.) This fact may indicate that individuals have difficulties in distinguishing between risk reductions involving small probabilities unless they are given some kind of reference point, as for example, when WTP is elicited for several risk reductions within a sample. In contrast to the Smith and Desvousges (1987) study, more recently Horowitz and Carson (1993) found strong support for the hypothesis that individuals do prefer to reduce risks for which the baseline risk is higher.

The reader should note that there is empirical evidence that people tend to overestimate small risks while tending to underestimate large risks (see Viscusi, 1992, Chap. 6, for details). However, this bias in risk perception does not necessarily mean that people are biased in their valuation of *changes* in risks. The empirical evidence in this respect is mixed, as is highlighted by the aforementioned and somewhat contradictory results obtained by Smith and Desvousges (1987) and Horowitz and Carson (1993), respectively.

The studies by both Jones-Lee et al. (1985) and Smith and Desvousges (1987) concerned mortality risks, but CV studies of morbidity risks are also starting to appear in the literature (see, for example, Berger et al., 1987; Evans and Viscusi, 1991). The contingent valuation method has also been used to evaluate the total economic losses from groundwater degradation; see pp. 113–16 of this chapter and Appendix 3 of Chapter 6 for details. There are also experiments with risk–risk trade-offs instead of WTP (see Viscusi *et al.*, 1991; and Krupnick and Cropper, 1992). This means that instead of investigating how much income people are prepared to give up in order to secure a risk reduction, the trade-off between a morbidity risk and a mortality risk is investigated. Individuals may for instance be indifferent between their initial situation and a project that reduces the risk of chronic bronchitis from 100/100 000 to 90/100 000 and increases the risk of automobile accident fatalities from 10/100 000 to 11/100 000. In this case the automobile accident death equivalent of chronic bronchitis is 0.1. The WTP per statistical life saved from automobile fatalities can then be used to compute the WTP per statistical case of chronic bronchitis prevented, to be used in a cost–benefit analysis. It should be noted that if the WTP approach and the risk–risk approach are equally valid and reliable, then it is more straightforward to directly elicit the WTP for the reduction in the morbidity risk. To compare the validity and reliability of these approaches to valuing reduced morbidity is an important issue for future research.

The CV method has so far been used mainly to value environmental health risks, but studies are also starting to appear in the healthcare field. For a review of such studies, the reader is referred to Johansson (1995).

The value of a statistical life

Comparisons between studies and methods are usually carried out in terms of the value per statistical life. It is, however, important to bear in mind that the value per statistical life can be expected to vary with the type of risk (e.g. voluntary versus involuntary), the initial risk level, the size of the risk change, age and income.

A number of reviews of the value per statistical life in the literature have been carried out; see, for example, Fisher *et al.* (1989), Jones-Lee (1996), Miller (1990) and Viscusi (1992, 1993). In the survey by Viscusi (1992, 1993), the value per statistical life varies between US$0.6 million and US$16.2 million in the surveyed labour market analyses of wage–risk trade-offs, between US$0.07 million and US$4.0 million in the studies of consumer markets, and between US$0.1 million and US$15.6 million in the CV studies (in December 1990 dollars). Viscusi (1992) noted that for labour market studies of wage–risk trade-offs 'most of the reasonable estimates of the value of life are clustered in the US$3 to US$7 million range' (Viscusi, 1992, p. 73). He also noted that this estimate conforms quite well with the results of the large-scale CV studies, whereas the results of the studies of consumer markets are well below this estimate. A similar conclusion was reached by Fisher *et al.* (1989), who stated that the newer CV and wage–risk studies yield similar results with respect to the value per statistical life. Fisher *et al.* concluded that 'The most defensible empirical results indicate a range for the value-per-statistical-life estimates of US$1.6 million to US$8.5 million (in 1986 dollars)' (Fisher *et al.* 1989, p. 96). Both Fisher *et al.* (1989) and Viscusi (1992) concluded that the estimates from the studies of consumer markets are biased downwards owing to the assumptions made in these studies and the problems of isolating the income–risk trade-off from confounding factors.

Miller (1990) in his review identified 65 studies of the value per statistical life, and eliminated 18 of those as unreliable. He then adjusted the value of the remaining 47 studies according to, for instance, risk perception. After the adjustments he found a mean value per statistical life of US$2.2 million, with $2.2 million for the labour market studies of wage–risk trade-offs, and US$2.5 million for the CV studies (in 1988 dollars). The adjustments and elimination of studies made by Miller were, however, to a large extent arbitrary and the review therefore exaggerated the similarity of results across different methods and studies.

The different methods discussed here are all associated with important advantages and disadvantages. The major problem of the indirect methods is to isolate the income–risk trade-off from confounding factors and to take into account the institutional restrictions of the observed markets. The area of application is also limited to those risks that are traded in markets, whereas the need for information about the value of health changes is greatest in the areas where no markets exist. The CV method overcomes these problems of the indirect methods, but its major disadvantage is, of course, that little is still known about to what extent answers to hypothetical questions mimic actual behaviour. Most studies based on actual behaviour of markets have been based on objective rather than subjective risks, which will bias the results to the extent that these risks diverge. In principle it should, however, be possible to adjust the results for this factor, to the extent that it is possible to measure the subjective risk levels. Although the CV methods usually ask the respondents to value a specified risk reduction, little is known about how such risk information is processed and to what extent the risk reductions are taken at face value or weighted by some prior belief.

An important issue is how to deal with altruism (i.e. that people are concerned about other people's welfare or safety) in cost–benefit analysis.

Indirect methods based on market behaviour probably do not capture altruistic values. The issue of altruism has to a large extent been neglected in empirical studies, with the exceptions of that of Jones-Lee *et al.* (1985), who found that the value per statistical life increased by about a third if (probably paternalistic or safety-oriented) altruism is included, and those of Viscusi *et al.* (1988) and Johannesson *et al.* (1996). The investigation of the role of altruism in the value of health changes is an important issue for future research.

Groundwater contamination

In this final section, we discuss briefly some aspects of the valuation of changes in groundwater quality. As is further discussed in Chapter 6, contamination of the groundwater may cause health effects as well as many other impacts on the community including non-use aspects such as concern for ecological resources, or fish and wildlife habitat. The averting expenditures method provides a possibility of capturing the use value of improved water quality. In response to a degradation of groundwater quality, individuals may respond by installing water filters or increase their demand for bottled water. As is illustrated by equations (5.3) and (5.5), one can use market prices to evaluate such averting expenditures. Examples of such studies of groundwater contamination are those by Abdalla (1990), Abdalla *et al.* (1992) and Harrington *et al.* (1989).

Abdalla (1990) measured household-level averting behaviours among 1596 households served by a public community water system in College Township in central Pennsylvania. The system relied exclusively on groundwater. In 1987 it was found that perchloroethylene (PCE), a volatile organic chemical, was present in the water supply (in 20–32 parts per billion (ppb) and averaged 25 ppb). Since no federal or state safety standard or maximum contaminant level existed, individuals were left to decide for themselves whether to drink or otherwise use the water. However, the state agency advised customers of potential health risks in statisical terms, for example expected cancer deaths in a population of a million, and provided examples of activities, for example cigarette smoking, with risks similar to drinking water containing perchloroethylene. According to a postal questionnaire, more than 75% of the households had made some adjustment to the contamination of the water supply. The actions included: (1) increased bottled water purchases among households buying water prior to the contamination, (2) bottled water purchases by new purchasers, (3) hauling water from alternative sources, (4) installing home water treatment systems, and (5) boiling water. In addition, about 35% adjusted their food and beverage purchases, for example increasing their purchases of ready-to-use fruit juices and soft drinks. An average of an extra US$126 was incurred over the six-month contamination period by the households estimated to have taken some averting action. The most significant predictors of the probability that a household would undertake avoidance actions were: the qualitative rating of perchloroethylene health risks, information received on avoidance practices, whether a pregnant woman was present in the household, and whether children under the age of 5 years were present in the household. On the other hand, the more information received on the health risks associated with perchloroethylene and the more trust a person had in state and local institutions, the less likely it was that the person did undertake avoidance actions.

Abdalla *et al.* (1992) considered averting behaviour among households in the borough of Perkasie in south-eastern Pennsylvania. In late 1987, a trichloroethylene (TCE), a volatile synthetic organic chemical, was detected in one of the borough's wells. Levels were as high as 35 ppb, exceeding the Environmental Protection Agency's maximum contaminant level of 5 ppb. According to the results reported by Abdalla *et al.* (1992), a household was more likely to take averting action if they had received information about trichloroethylene, and rated the cancer risks associated with the levels of trichloroethylene in their water to be relatively high, and children in the ages 3 to 17 years were present in the household. The average weekly increase in averting expenditures per household which undertook averting actions in response to the contamination was US$0.40 during the 88 weeks covered by the study.

Harrington *et al.* (1989) estimated the costs of an increase in the incidence of giardiasis, currently the most common waterborne disease in the USA, in a number of small communities near Wilkes-Barre, Pennsylvania, during the late Fall of 1983. Although seldom fatal, giardiasis, caused by the protozoan parasite *Giardia lamblia*, can be an unpleasant and temporarily debilitating diarrhoeal disease. Untreated, giardiasis often develops into chronic infection, characterised by recurrent periods of acute illness lasting several days. Interviewed people chose a wide variety of strategies to ensure a safe drinking-water supply. About half of the households either hauled water or boiled water, but not both. Only 2% relied on bottled water alone, and no household in the sample installed a filtration system. The 'best estimate' of averting losses to the average household range from US$485 to US$1540 or from US$1.13 to US$3.59 per person per day for the duration of the outbreak. The quite wide range is due to different assumptions about the value of the time spent in averting activities (throughout, using the after-tax wage rate for the employed, but varying the cost for the unemployed, homemakers, and retired persons).

There are also attempts to use property values to evaluate losses from groundwater degradation (see, for example, Malone and Barrows, 1990). The idea is that contamination of wells in an area, *ceteris paribus*, will cause property values in that area to be lower than property values in uncontaminated areas. Malone and Barrows (1990) analysed property transactions of residences with nitrate-contaminated wells in Portage County, Wisconsin, but found no statistically significant effect of nitrate contamination on property values. However, it was concluded that contamination did create household-level costs, such as sellers' remediation or treatment of the problem prior to the sale.

Market prices can be used to evaluate losses of use values from groundwater degradation. However, people may be concerned about the well-being of others as well as about ecological resources. In other words, in addition to use values there may be non-use values such as (inter-household) altruistic values and pure existence values associated with a changed groundwater quality. Such non-use values are not covered in market prices, in general. For this reason, some authors have used the contingent valuation method to evaluate the total economic losses from groundwater degradation. In Cape Code, MA Edwards (1988) conducted a contingent valuation study of residents' WTP to prevent uncertain future nitrate contamination of a portable groundwater supply. The average annual WTP per household range between US$500 and US$2500 to preserve future use of the aquifer, and altruistic concerns for future generations had a strong impact on the magnitude of the WTP. Other contingent valuation studies of nitrate contamination of the water supply include those by Hanley (1989) and Silvander (1991) (see also the brief review of CV studies in Appendix 3 of Chapter 6). Mitchell and Carson (1986) focused on drinking-water quality. In the early 1980s, local water companies in southern Illinois had difficulty in meeting the US standard for trihalomethane (THM), a substance considered to pose potential carcinogenic risk to humans. A sample of individuals living in this area were asked to state their WTP for a reduction of the trihalomethane level from some hypothetical level to the standard. The corresponding mortality risk reductions were communicated with the aid of risk ladders. The largest and smallest risk reductions were a decrease in annual mortality risk from 9.5 and 0.61 per 100 000 persons, respectively, to 0.57 per 100 000. The smallest risk reduction is comparable with avoiding the risk of being struck and killed by lightning. The estimated average annual WTP per household was about US$52 and US$3.5 for the largest and smallest risk reduction, respectively. In this study, therefore, people are sensitive to the magnitude of the risk reduction as well as able to evaluate even very small risk reductions. Chapter 6, which deals with groundwater quality in Milan, uses the contingent valuation method to assess the value of improved water quality.

Appendix

In order to arrive at the indirect utility function in equation (5.4), let the individual maximise $U[\pi(X), x]$ subject to his budget constraint $y = px + PX$. First-order conditions for an interior solution read:

$$\delta U(.) / \delta x = \lambda p \tag{A5.1}$$

$$(\delta U(.) / \delta \pi)(\delta \pi / \delta X) = \lambda P$$

$$y = px + PX$$

where λ is the Lagrange multiplier associated with the budget constraint. Solving (A5.1) yields demand functions of the form $x = x(p, P, y)$ and $X = X(p, P, y)$. Substitution of these into the direct utility function yields equation (5.4). Alternatively, totally differentiating the direct utility function using (A5.1) and a totally differentiated budget constraint establishes the fact that utility is a function of prices and income.

In the case of intrahousehold altruism, the relevant first-order condition reads:

$$\Sigma_m(\delta U(.) / \delta \pi_m)(\delta \pi_m / \delta X) = \lambda P \tag{A5.2}$$

Thus, the sum across household members of their marginal willingness to pay is set equal to the market price of the health good. In order to arrive at the definition of the value of statistical life, assume a two-states world, and also that the expected utility of individual h can be written as follows:

$$V^{hE} = \pi_{h1} V^h(p, y_h, z_1) \qquad \forall h \tag{A5.3}$$

where π_{h1} is the survival probability of individual h, z_1 denotes being alive and the utility of not being alive, i.e. $V^h = V^h(p, y_h, z_0)$, where $z = z_0$ means not being alive, is set equal to zero. Differentiating (A5.3) with respect to π_{h1} and adjusting income so as to keep expected utility constant yields, after straightforward calculations:

$$\mathrm{dCV}_h = V^h(p, y_h, z_1)\mathrm{d}\pi_{h1} / V^h_y(.) = \mathrm{MRS}_h \mathrm{d}\pi_{h1} \qquad \forall h \qquad (A5.4)$$

where dCV_h is the marginal non-contingent compensating variation, V^h_y is the marginal utility of income if alive, and MRS_h denotes the marginal rate of substitution between income and risk. If $\mathrm{d}\pi_{h1} = b/H$ for all h, summing across individuals yields:

$$\Sigma_h \mathrm{dCV}_h / b = \Sigma_h \mathrm{MRS}_h / H \qquad (A5.5)$$

since $\Sigma_h \mathrm{dCV}_h = \Sigma_h \mathrm{MRS}_h \mathrm{d}\pi_{h1} = \Sigma_h \mathrm{MRS}_h b/H$. The right-hand side expression in (A5.5) yields the average value of a marginal reduction in the probability of death. The left-hand side expression yields the aggregate WTP per life saved by the considered measure, i.e. what is known as the value of statistical life.

The reader is invited to differentiate (A5.3) with respect to π_{h1} and calculate the associated change in expected utility. Next, adjust income through an amount dEV_h so as to obtain the same change in expected utility. The amount dEV_h represents the compensation the individual needs in order to be as well off as with an increase in the survival probability. Apparently, it must be the case that $\mathrm{dEV}_h = \mathrm{dCV}_h$. The value of statistical life in (A5.5) is thus the same regardless of whether we use the (marginal) CV or the EV measure.

Notes

1 We are grateful to a referee and the editors for comments on an earlier version of this chapter. This research was financially supported by the Swedish Council for Social Research.
2 For clarity, we use 'he' as a generic term for male and female throughout this chapter.
3 Pages 105–13 draw heavily on Johannesson *et al.* (1995).

References

Abdulla, C. W. (1990). Measuring economic losses from ground water contamination: an investigation of household avoidance costs. *Water Resources Bulletin*, **26**, 451–63.
Abdulla, C. W., Roach, B. A. and Epp, D. J. (1992). Valuing environmental quality changes using averting expenditures: an application to groundwater contamination. *Land Economics*, **68**, 163–9.
Acton, J. P. (1973). Evaluating public programs to save lives: the case of heart attacks. *RAND Report* R-950-RC, Santa Monica, CA.
Åkerman, J., Johnson, F. R. and Bergman, L. (1991). Paying for safety: voluntary reduction of residential radon risks. *Land Economics*, **67**, 435–46.

Berger, M. C., Blomquist, G. C., Kenkel, D. and Tolley, G. S. (1987). Valuing changes in health risks: a comparison of alternative measures. *Southern Economic Journal*, **53**, 967–84.

Blomquist, G. (1979). Value of life savings: implications of consumption activity. *Journal of Political Economy*, **87**, 540–58.

Crocker, T. D., Forster, B. A. and Shogren, J. F. (1991). Valuing potential groundwater protection benefits. *Water Resources Research*, **27**, 1–6.

Dardis, R. (1980). The value of a life: new evidence from the marketplace. *American Economic Review*, **70**, 1077–82.

Edwards, S. F. (1988). Option prices for groundwater protection. *Journal of Environmental Economics and Management*, **15**, 475–87.

Epple, D. (1987). Hedonic prices and implicit markets: estimating demand and supply functions for differentiated products. *Journal of Political Economy*, **95**, 59–80.

Evans, W. and Viscusi, W. K. (1991). Estimation of state-dependent utility functions using survey data. *Review of Economics and Statistics*, **73**, 94–104.

Fisher, A., Chestnut, L. G. and Violette, D. M. (1989). The value of reducing risks of death: a note on new evidence. *Journal of Policy Analysis and Management*, **8**, 88–100.

French, M. T. and Kendall, D. L. (1992). The value of job safety for railroad workers. *Journal of Risk and Uncertainty*, **5**, 175–85.

Hanley, N. (1989). Problems in valuing environmental improvements resulting from agricultural policy changes: the case of nitrate pollution. Unpublished report given in AEEA seminar on economic aspects of environmemtal regulations in agriculture. Thune University of Agriculture, Copenhagen.

Harrington, W., Krupnick, A. J. and Spofford, W. O. Jr (1989). The economic losses of waterborne disease outbreak. *Journal of Urban Economics*, **25**, 116–37.

Horowitz, J. K. and Carson, R. T. (1993). Baseline risk and preference for reductions in risk-to-life. *Risk Analysis*, **13**, 457–62.

Hwang, H., Reed, W. R. and Hubbard, C. (1992). Compensating wage differentials and unobserved productivity. *Journal of Political Economy*, **100**, 835–58.

Johannesson, M., Johansson, P.-O., Jönsson, B. and Söderquist, T. (1995). Valuing changes in health. Theoretical and empirical issues. In *Current Issues in Environmental Economics*, ed. P.-O. Johansson, B. Kriström and K. G. Maler. Manchester University Press, Manchester.

Johannesson, M., Johansson, P.-O. and O'Conor, R. (1996). The value of private safety versus the value of public safety. *Journal of Risk and Uncertainty*, **13**, 263–75.

Johansson, P.-O. (1994). Altruism and the value of statistical life: empirical implications. *Journal of Health Economics*, **13**, 111–18.

Johansson, P.-O. (1995). *Evaluating Health Risks. An Economic Approach*. Cambridge University Press, Cambridge.

Jones-Lee, M. W. (1976). *The Value of Life: An Economic Analysis*. Martin Robertson, London.

Jones-Lee, M. W. (1991). Altruism and the value of other people's safety. *Journal of Risk and Uncertainty*, **4**, 213–19.

Jones-Lee, M. W. (1992). Paternalistic altruism and the value of a statistical life. *Economic Journal*, **102**, 80–90.

Jones-Lee, M. W. (1996). Summaries of selected publications on contingent valuations and other direct preference-elicitation methods in the fields of health and safety. Appendix A to a report to the Health & Safety Executive.

Jones-Lee, M. W., Hammerton, M. and Philips, P. R. (1985). The value of safety: results of a national sample survey. *Economic Journal*, **95**, 49–72.

Kniesner, T. J. and Leeth, J. D. (1991). Compensating wage differentials for fatal injury risk in Australia, Japan and the United States. *Journal of Risk and Uncertainty*, **4**, 75–90.

Kohlhase, J. E. (1991). The impact of toxic waste sites on housing values. *Journal of Urban Economics*, **30**, 1–26.

Krupnick, A. J. and Cropper, M. L. (1992). The effect of information on health risk valuations. *Journal of Risk and Uncertainty*, **5**, 29–48.

Malone, P. and Barrows, R. (1990). Groundwater pollution's effects on residential property values. *Journal of Soil and Water Conservation*, **45**, 180–3.

Miller, T. R. (1990). The plausible range for the value of life: red herrings among the mackerel. *Journal of Forensic Economics*, **3**, 17–40.

Mitchell, R. C. and Carson, R. T. (1986). Valuing drinking water risk reductions using the contingent valuation method: a methodological study of risks from THM and giardia. Draft report to the U.S. Environmental Protection Agency. Resources for the Future, Washington, DC.

Portney, P. R. (1981). Housing prices, health effects, and valuing reductions in risk of death. *Journal of Environmental Economics and Management*, **8**, 72–8.

Rosen, S. (1986). The theory of equalizing differences. In *Handbook of Labor Economics*, ed. O. Ashenfelter and R. Layard, pp. 641–92. Elsevier Science Publishers, Amsterdam.

Rosen, S. (1988). The value of changes in life expectancy. *Journal of Risk and Uncertainty*, **1**, 285–304.

Shogren, J. F. and Crocker, T. D. (1991). Risk, self-protection, and *ex ante* economic value. *Journal of Environmental Economics and Management*, **20**, 1–15.

Silvander, U. (1991). The willingness to pay for angling and groundwater in Sweden. Dissertation 2, Dept. of Economics, Swedish University of Agricultural Sciences. Uppsala.

Smith, V. K. and Desvousges, W. H. (1987). An empirical analysis of the economic value of risk changes. *Journal of Political Economy*, **95**, 89–115.

Söderqvist, T. (1991). Measuring the value of reduced health risks: the hedonic price technique applied on the case of radon radiation. Unpublished research report, The Economic Research Institute at the Stockholm School of Economics.

Viscusi, W. K. (1992). Fatal Tradeoffs. Public & Private Responsibilities for Risk. Oxford University Press, New York.

Viscusi, W. K. (1993). The value of risks to life and health. *Journal of Economic Literature*, **31**, 1912–46.

Viscusi, W. K., Magat, W. A. and Forrest, A. (1988). Altruistic and private valuations of risk reduction. *Journal of Policy Analysis and Management*, **7**, 227–45.

Viscusi, W. K., Magat, W. A. and Huber, J. (1991). Pricing environmental health risks: survey assessments of risk–risk and risk–dollar trade-offs for chronic bronchitis. *Journal of Environmental Economics and Management*, **21**, 32–51.

6   On estimating the benefits of groundwater protection: a contingent valuation study in Milan

Jane Press and Tore Söderqvist[1]

Introduction

In Italy, atrazine has been used extensively in maize cultivation since the early 1960s. One of the advantages of using atrazine in maize cultivation is flexibility in the timing of the application. Unfortunately, however, its chemical properties make it a likely source of groundwater contamination, especially in areas with permeable soils. In some areas of the Po Valley in northern Italy, where maize cultivation is extensive and groundwater the dominant source of drinking water, chemical monitoring in the 1980s revealed that concentrations of atrazine exceeded 0.1 µg/l i.e. the maximum admissible concentration set by the Directive on Drinking Water Quality of 15 July 1980 (see Vighi and Zanin, 1994, for examples of monitoring results). Following the implementation of local restrictions in 1986, a nationwide ban on the sale and use of atrazine was introduced in 1990.

An earlier pilot study was conducted to address the question of whether the EC standard, with specific reference to atrazine in Italy, could be justified in terms of social efficiency; that is, is 0.1 µg/l a socially efficient contamination level (Bergman and Pugh, 1997)? Broadly speaking, this question can be answered in the affirmative if 0.1 µg/l is the concentration of atrazine in drinking water for which the marginal cost of reducing the concentration is equal to the marginal benefit of a reduced concentration (see Söderqvist *et al.*, 1995, for a basic introduction to these issues). We confine ourselves here to noting that reduction costs may, *inter alia*, be due to farmers having to turn to more costly weed eradication methods, and that reduced health risks may be one important constituent of reduction benefits. Of course, the benefits may also involve reduced risks of environmental damage per se, but since the EC standard is for drinking water, not groundwater, the pilot study focused on human health issues.

122   J. Press and T. Söderqvist

It soon became evident, however, that lack of information on the benefits of reducing atrazine contamination precluded an analysis approaching social efficiency. This was not least due to the fact that the health risks caused by observed atrazine levels in Italian groundwater reserves are virtually zero (Vighi and Zanin, 1994). Other, less tangible benefits of reducing atrazine concentration may nevertheless exist. Even if objective health risks are considered to be almost zero, people may perceive the matter differently and be willing to give up some wealth for accomplishing decreased concentrations of pesticides in groundwater. In an economic analysis, such willingness has to be taken into account. However, the information available on such benefits was found to be very limited, since it is not usually expressed in monetary terms. For the pilot study, this problem had the consequence that the economic analysis was restricted to a study of cost-effectiveness rather than one of economic efficiency. More precisely, Söderqvist (1994) studied the costs of two ways of accomplishing the EU standard: via the ban actually introduced in Italy and, alternatively, via water treatment techniques.

The lack of information on benefits is disturbing. It precludes conclusions on efficiency and consequently on the appropriate degree of regulation. For example, the accumulation tax on pesticides suggested by Swanson in Chapter 9 of this volume should reflect the value of the groundwater resource. A benefit study was therefore initiated and this chapter reports the design and results of this study. In order to build upon knowledge gained by the atrazine pilot study, the benefit study was applied to the situation in northern Italy.

This chapter is structured as follows. The next section describes available benefit estimation methods and the choice of estimation method for this case study. The third section is devoted to some issues related to the application of the selected methodology – the contingent valuation method (CVM) – in the Italian context. Design of the CVM study constitutes another section. The results and empirical methods of the study are then presented, followed by a discussion.

Benefit estimation method

Introduction

Groundwater is the main source of drinking water in northern Italy. The city of Milan, in particular, relies entirely on groundwater reserves for municipal water supplies. This suggests that satisfactory groundwater quality is of considerable importance to the inhabitants of this part of Italy. The use of groundwater as a drinking-water source implies that an increased concentration of potential pollutants[2], such as herbicides, in groundwater may compromise groundwater quality, leading to a corresponding reduction in drinking-water quality, if no mitigating action is taken. Such actions include dilution of contaminated supplies with cleaner water from other sources, or various types of chemical treatment to remove dangerous substances. In addition, a change in groundwater quality may be of importance to people in other ways. For example, groundwater is used for irrigation in many areas in northern Italy and any contaminants present may cause damage to living organisms that are of relevance to human well-being (either directly via incorporation into the foodchain or indirectly as a consequence of ecosystem imbalances).

Despite the importance of satisfactory groundwater quality to people, it does not have any market price per se. This is also true for many other goods and services supplied by the environment. The lack of market prices is due mainly to the public nature of such resources: they often involve at least some element of non-excludability (i.e. no one can be excluded from consumption; it is a common resource) and/or non-rivalry (i.e. one individual's consumption does not affect the amount of the good available to others). Conversely, for goods and services that are freely traded and priced in a market, conventional economic theory allows an interpretation of the market price as a measure of the economic value of the good in question. The market price is therefore a natural basis for estimations of the change in welfare that would arise from a change in the supply of that good. However, when a market price does not exist, often referred to as *market failure*, the determination of the economic value becomes more complex – as in the case of groundwater.

Traditional welfare economics[3] suggests that economic value can be measured as individuals' willingness to make economic sacrifices or to accept economic compensations that correspond to changes in the supply of the non-market good being considered. In order to make these ideas more clear, let us follow Johansson (1993, pp. 24–6) and describe an individual's utility (i.e. the conventional economic measure of well-being) in very general terms as:

$$u = U(x, z) \tag{6.1}$$

where u denotes utility; z is the groundwater quality that the individual enjoys; x is a vector $(x_1, \ldots, x_n)$ describing the individual's consumption of all goods and services other than z; and $U(.)$ is a function mapping combinations of x and z into utility.

Since groundwater quality per se does not have any market price, the budget restriction of an individual is simply:

$$px = y \tag{6.2}$$

where p is a vector $(p_1, \ldots, p_n)$ of prices of goods and services other than z, and y is income.

The result of a maximisation of utility subject to the budget restriction can be described by an indirect utility function:

$$v = U[X(p, y, z), z] = V(p, y, z) \tag{6.3}$$

where v is indirect utility; $X(.)$ is a vector of demand functions $[X_1(.), \ldots, X_n(.)]$ for all goods and services other than z; and $V(.)$ is the indirect utility function.

Now consider a change in groundwater quality from an initial level z^0 to z^1. For simplicity, let us assume that this change does not affect p or y. The two most common measures of the welfare change due to the change in z is the equivalent variation (EV) and the compensating variation (CV) (see, for example, Ng, 1979, for a discussion of some less commonly used welfare change measures). EV and CV can be defined implicitly as:

$$V(p, y \pm \text{EV}, z^0) = V(p, y, z^1) \tag{6.4a}$$
$$V(p, y \pm \text{CV}, z^1) = V(p, y, z^0) \tag{6.4b}$$

The change in groundwater quality can either be an improvement or a deterioration. Let us first consider the case when $z^1 > z^0$, i.e. an improvement, so that $V(p, y, z^1) > V(p, y, z^0)$. In this case, equation (4a) implies that EV is the minimum money compensation that the individual is willing to accept in order to voluntarily agree to a *status quo*, i.e. to remain in situation 0. In equation (4*b*), the matter is seen from a slightly different perspective: CV is the maximum amount of money that the individual is willing to pay in order to attain the change from situation 0 to 1.

We also have the case when $z^1 < z^0$, i.e. a deterioration of the groundwater quality, so that $V(\boldsymbol{p}, y, z^1) < V(\boldsymbol{p}, y, z^0)$. In this case, equation (4*a*) implies that EV is the maximum amount that the individual is willing to pay in order to avoid the change from situation 0 to 1. This amount can be interpreted as the maximum willingness to pay for groundwater quality protection. Equation (4*b*) implies that CV is the minimum monetary compensation that the individual is willing to accept in order voluntarily to agree to the change from situation 0 to 1.

These definitions of EV and CV suggest that a welfare change can be measured in terms of an individual's willingness to pay (WTP) or willingness to accept compensation (WTA). It is also clear that it depends on the particular welfare change under consideration whether the WTP (or WTA) measures an EV or a CV. The correspondence between EV/CV and WTP/WTA is acknowledged by the methods that have been developed within the field of environmental economics for the estimation of the size of the welfare change due to a changed provision of an unpriced environmental service. These methods – often termed benefit estimation methods – are presented briefly in the next subsection. A typical application involves the collection of data on the WTP for a particular change of a sample of individuals belonging to the affected population. A central tendency measure of WTP is then estimated. In a traditional cost–benefit analysis perspective, the relevant measure is *mean* WTP, from which an aggregate WTP for the population in question can be estimated. An estimate of *median* WTP may in other instances be of relevance.[4] We focus on estimates of mean WTP in this chapter because of the social efficiency issue that prompted the benefit study.

Note finally that the methods presented below all share the individualistic perspective (i.e. the WTP/WTA perspective) of welfare economics. Other methods, for example the so-called human capital approach, will not be considered here (see, for example, Freeman, 1993, pp. 322–5), on this approach).

Approaches to benefit estimation

Following the methods of Freeman (1993, pp. 23–6) and Mitchell and Carson (1989, pp. 74–87), two features can be used for a classification of benefit estimation methods. The first feature is whether a method employs data on individuals' *observed* or *hypothetical* behaviour. The second feature is whether a WTP (or WTA) estimate is obtained in a *direct* or *indirect* way.

Observed behaviour

The use of simulated markets is an Observed/Direct method, whereby experimental markets are set up by researchers. A sample of individuals is invited to actually trade the normally unpriced environmental service among themselves (see Bishop and Heberlein, 1979, for an example). More commonly used estimation methods are found in the Observed/Indirect class. These rely on the potential existence of linkages between the environmental service and one or several market goods, such as some kind of substitute or complementary relationship. Given such a relationship, people's WTP (or WTA) for a change in the level of provision of the environmental good/service may be inferred by studying demand for the related market good(s). Three examples of Observed/Indirect estimation methods are: the 'hedonic price' method, which studies the influence of a good's characteristics on its market price (for example the relationship between ambient air quality and housing prices) the 'travel cost' method, which focuses on the costs of travel connected with use of recreational areas; and the 'defensive expenditure' technique, which studies the demand for a market good that at least to some extent substitutes for the environmental service (for example the demand for water filters that give some protection against falling water quality). (See, for example, Braden and Kolstad, 1991 and Freeman, 1993, for presentations of Observed/Indirect methods.)

Hypothetical behaviour

The contingent ranking method belongs to the Hypothetical/Indirect category. When this method is used, a sample of people are asked to rank situations that differ with respect to the quantity available of the non-market good and the associated payment requirements. WTP estimates can then be inferred from the ranking (see Smith and Desvousges, 1986, for an example). Finally, there is the Hypothetical/Direct class. This includes the contingent valuation method (CVM) whereby a change in the level of provision of the environmental good or service in question is clearly described to a sample of individuals, using a standard survey instrument such as mail questionnaires, telephone interviews or face-to-face interviews, and questions are then posed about the WTP of respondents to effect the change. Payment questions may be set at an individual or household level. Actual payments, however, do not take place. (See Mitchell and Carson, 1989, for a full exposition of this method.)

Selection of benefit estimation method

The various Observed/Indirect methods and the CVM have been applied most extensively by environmental economists in the last few decades. It was primarily these methods that were considered for exploring the benefits of protecting Milan's groundwater resources. Our prior conceptions of the determinants of the benefits of protecting Milan's groundwater quality were of importance for the choice of an estimation method. We believed that the main reason that people might value the quality of Milan's groundwater is its implications for potable (tap) water quality. Tapwater is used both as drinking water and for myriad other purposes which means that groundwater quality impacts upon several different activities ranging from drinking-water-related health risks to the build-up of scale residues (calcium carbonate deposits) in household appliances such as washing machines. Moreover, it cannot be precluded that some people also value the quality of Milan's groundwater because of reasons other than those related to their use of tap water. Here we enter the domain of generally termed 'non-use values'. McClelland *et al.* (1992, pp. 3–4) identified three types of non-use value of groundwater, all of which may apply in the case of Milan:

(1) *Altruistic value*: the value that an individual places on other persons having protected groundwater reserves today.
(2) *Bequest value*: the value that the current generation places on protected groundwater reserves for the benefit of future generations.
(3) *Existence value*: the value that an individual places on simply knowing – independently of any conceived present or future use – that groundwater resources are protected.

Another aspect is the indirect use value of groundwater quality as a type of barometer for pollution flows through the wider environmental system – industrial effluents, agricultural and urban runoff and traces of atmospheric pollution dissolved in rainwater ultimately seep into underground waters, where they may remain, inert, for decades or else percolate into surface water systems, depending on the geological make-up of the acquifer system. With these concepts to describe the various benefits of defending groundwater quality in mind, let us turn to the question of whether the size of these benefits can be revealed by market data. In other words, is an application of an Observed/Indirect method possible in the case of Milan?

A first important observation is that tap water has a price, in the form of municipal water tariffs which are set by the Milan water authorities. However, these tariffs are not any result of market transactions, but are decided by the municipal authority, which has a supply monopoly. Apart from the fact that clean water is an essential, life-supporting good and everyone in society is entitled to their fair share, this means that the population's water expenditure, in the form of tariffs, cannot be considered as any indication of the benefits society derives from water supply services and, by extrapolation, groundwater quality.

Two other real markets exist, however, which potentially correlate more closely to water quality preferences: *domestic water filters* and *mineral water*. Some households buy tap water filters in order to improve water quality as these appliances (ostensibly) remove various chemical contaminants and/or reduce water hardness. Such filters, at least partly, substitute for satisfactory groundwater quality. Motives for installing a water filter may, however, relate to natural groundwater characteristics such as hardness (calcium carbonate concentration), which affects mainly taste and scale residues, as well as to characteristics deriving from human activities, such as chemical residues that present potential health risks. It is interest in the latter and policies related to controlling human impacts that constitute the background to this study. This means that data on domestic water filter demand, without accurate information on the reasons for their installation, could be difficult to interpret from a policy point of view. Note also that filter market data can only give some information on the direct *use* benefits of protecting groundwater to assure tap water quality, while it is likely to overlook the potential *non-use values* of protecting groundwater reserves.

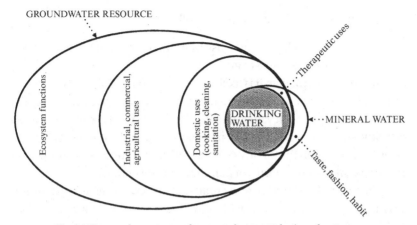

Fig. 6.1. Demand parameters for groundwater and mineral water.

As to the market for mineral water, the most obvious drinking-water alternative, the matter is confused by the fact that mineral water is only partly used as a direct substitute for tap water. For example, mineral water may be consumed instead of wine or other alcoholic drinks when dining or for specific therapeutic reasons. Moreover, in cases when mineral water is indeed used as a substitute for drinking water, the problem just described for the case of water filters again arises: the existence of several different water quality parameters (taste, health risks, etc.) implies that it is not self-evident why a person chooses to substitute mineral water for tap water. These substitutability issues are illustrated in Figure 6.1.

Considering their various demand-related characteristics, the two 'goods' display a relatively small overlap which is confined to the drinking-water segment. Market studies have clearly shown that mineral water demand, especially in recent years, is strongly driven by lifestyle trends and specific marketing strategies, as well as distinct therapeutic properties (Databank, 1993). Mineral water quality, moreover, is governed by a set of regulations separate and distinct from those of potable municipal water (De Bernardi *et al.*, 1993). In the case of groundwater, non-drinking features account for an even larger share of the total resource value in terms of additional uses (cooking, cleaning, washing plus agricultural, commercial and industrial activities) as well as non-use characteristics and wider ecosystem functions. The contribution made by this category of benefits, as part of the full economic value of groundwater resources, would be overlooked by any valuation of water quality based on mineral water demand.

The contingent valuation approach

Clearly, the applicability of the Observed/Direct approach using market data to estimate the benefits of groundwater protection measures is somewhat limited. In particular, it is likely to exclude non-use values (cf. Larson, 1992). The contingent valuation method, however, has the potential directly to consider such values and it was therefore selected for the Milan case study. An application of this method also permits the researchers to focus specifically on the quality characteristics that are most relevant from a policy point of view, in terms of pollution control options.

The machinations of environmental debates have revealed that awareness lies at the root of concern, which, in turn, is instrumental in the investment of effort or attracting support for environmental initiatives. The execution of a CVM survey therefore has to tread a careful path to reveal concern and latent willingness to dedicate economic resources to environmental protection without influencing respondents' prevailing attitudes. In other words, the survey is not intended as an awareness-raising instrument, yet respondents must be empowered by the provision of some decision-making support information in order to mimic a real market situation as closely as possible.

The CVM is also a controversial tool because of its use of data on people's hypothetical market behaviour, i.e. decisions are contingent on there being a real market for the good or service being valued (see, for example, Diamond and Hausman, 1994, Hanemann, 1994, and Portney, 1994, for some recent reports on the CVM debate). These controversies call for circumspect applications of the CVM, an approach which takes account of the potential shortfalls in the methodology, compensating where possible in the preparatory and interpretative phases of the study. Precautions taken and specific details of the experiment design are described later.

A key element of this integrated approach to the valuation issue was the parallel study of demand for water filters and mineral water, in order to have a benchmark for the benefit estimates resulting from the CVM application. Moreover, since only a very few CVM studies have been undertaken in Italy, our study also faced the problem of exploring new ground as regards the choice and acceptance of the survey instrument. A thorough understanding of the water quality situation in Milan was also necessary for the design of an appropriate valuation scenario to be communicated to respondents. In the next section, we therefore turn to a background description of the water quality situation in Milan and issues related to the application of the CVM in the Italian context.

Italian context for the CVM study

The first thing to note is the relatively immature tradition of public opinion surveys in Italy, compared to the USA in particular. The work presented here is therefore less an attempt to precisely measure the full economic value of groundwater protection and more a case of exploring peoples' priorities, understanding and perceptions of water quality.

Furthermore, the sparse experience of the CVM that exists here has focused upon valuing nature reserves and similar entities characterised by *recreational* use value along with the various ecological and moral components of non-use value. The valuation issue in the case of groundwater reserves, in contrast, is more intricate: a number of essential features (drinking water, input to production activities) are (1) mixed with the non-use values of the good and (2) partially substitutable (mineral water for drinking water), as was outlined above.

The study location: Milan

The city of Milan, with a population of 1.6 million, is capital of the Milan province, part of the Lombardy region in northern Italy. It is the commercial and industrial capital of the country, surrounded by major industrial activities as well as intensive agriculture.

Metropolitan Milan was chosen for the study since its municipal water supply is exclusively sourced from natural acquifers lying below the city so that groundwater and drinking water policy issues overlap considerably. From an analytical perspective, however, the spatial dislocation between the various uses of the water resource, not ignoring water's function as a universal solvent and hence vector for by-products of economic activities (agricultural runoff, industrial effluents), must be noted. In other words, groundwater contamination in the city itself can be traced primarily to industry and agriculture located in the hinterland but still within the aquifer infiltration zone. The burden of surrounding activities on ambient groundwater quality and the disproportionately high consumption of public supplies by the capital of the province are further detailed in

Appendix 1 of this chapter. These activities are vital to life in the city, none the less, providing food, other goods and employment to its inhabitants. By confining the survey to the urban population, however, we avoided any negative WTP that can be expected among farmers, for example, who would seek financial compensation for production losses because of having to reduce their use of agrochemicals in order to meet the environmental objective.

Municipal water supply in Milan

The municipal water supply is the same as tap water in the case of households, also referred to here as drinking water. Currently, Milan's water narrowly meets the EU standards as set by the 1980 European Commission Directive on Drinking Water Quality (80/778/EEC) and subsequently adopted in Italian law (DPR 236/88). The European Directive contains certain guidelines as well as maximum permissible concentrations for 71 parameters. The Italian interpretation includes conditions under which regional administrations can issue limited exemptions to water authorities regarding certain standards, in view of the extensive remedial measures and investment pressures created by the new standards (Funari *et al.*, 1993). This facility was taken up by Milan's water authority in the case of chlorinated organics (solvent derived).

Due to the large groundwater reserves in the natural aquifers underlying the city, drinking-water standards have traditionally been met by closing the most heavily contaminated wells and mixing waters from wells of differing contamination levels before delivery to the water supply network. This was the main water quality management system adopted in Milan until spring 1994, when a number of activated carbon filters had to be installed in order to abate solvent levels to within legal limits.

Alongside institutional characteristics (see Appendix 1), this rather unintensive water management policy has permitted water tariffs to remain among the lowest in the country, which, in turn, has one of the lowest average water rates in Europe. The European average is around ITL1500/million^3 while in Milan it is six times lower at ITL250/million^3 (a mere ITL0.25/litre).[5] This accounts for an extremely low awareness of the 'cost' of drinking water among consumers, presenting significant problems in the execution of the CVM survey, since respondents, having little idea of what they currently pay for water services, have a correspondingly scant reference point from which to begin valuing natural water resources (at least in terms of their potential use value and irrespective of the theoretical arguments that municipal water rates are artificially set rather than via a real market mechanism).

A second confounding factor relating to Milan's water supply in the context of the CVM study is the prevailing confusion between water quality and its organoleptic properties. Due to natural features of the rock substrate constituting the aquifers, the calcium carbonate content, or hardness, of water supplies varies dramatically across the city, with corresponding changes in taste and scale residue deposit rates. As the public debate about health threats in drinking water becomes more acute, such features are mistakenly taken to represent traces of artificial contamination. As a consequence, people may exaggerate their WTP for groundwater protection, thinking that the current situation is worse than reality.

Water consumption behaviour

When a hypothetical valuation methodology is used, expressed WTP needs to be interpreted in the light of manifest behaviour as regards the resource to be valued.[6] In the case of groundwater protection, it is useful to have background knowledge of the population's actual use of municipal supplies as a source of clean, safe, drinking water. In other words, parameters affecting at least the use value component of the full economic value of groundwater.

Italy, and particularly northern Italy, is characterised by the highest per capita mineral water consumption levels in Europe (Databank, 1993). At first sight, this may seem to be a strong indication of people's preference for pristine drinking water and a measure of the extent to which they are prepared to pay to avoid perceived health risks of contaminants and uncertain quality of municipal water supplies. As mentioned on p. 129, however, mineral water demand is driven by a number of factors, many of which are completely unrelated to water quality concerns. High mineral water consumption levels can thus be taken to mean that many respondents are not directly exposed to tap water as a source of drinking water, albeit that the quality issues remain important in terms of cooking and cleaning uses. A secondary, but related issue, is the strong taste of tap water in certain areas of Milan, due to hardness, which makes some people prefer mineral water. This natural feature is often confused with chemical contamination, as mentioned in the previous section.

Study design

Survey instrument

The complexity of the valuation issue, implying the need for a questionnaire which covers many aspects of water demand, combined with unfamiliarity regarding this type of research in Italy, meant that the survey needed to be executed either via personal interviews or by mail, to allow for the large volume of information exchange. General survey experience in Italy reveals that the face-to-face mode is the most efficient; mail surveys typically have a response rate of well under 5%, even in the professional/corporate sector, according to local market research professionals.

The final survey was executed in collaboration with the Dipartimento di Sociologia at the Università di Milano, once a final version of the questionnaire had been developed to the pilot stage (as described on pp. 139–41). This research group had rare experience of surveys relating to environmental issues, having worked on studies concerning the siting of waste disposal facilities and the planning of a new high-speed railway line.

Valuation issue

While previous sections addressed the issue of estimating the economic value of groundwater protection, execution of a CVM survey requires a more precise specification of the entity in order to approximate the marketplace as effectively as possible, such that respondents are asked to make a type of purchasing decision about a clearly defined, if not totally familiar good. Significant time and effort was therefore dedicated to understanding the target population's current perception of the resource and its key characteristics in order to devise a readily understandable and appropriate *scenario description* and *valuation question*. These represent the good/service on offer in the marketplace and its purchasing conditions, respectively.

Difficulties include the multiplicity and intangibility of certain values that potentially contribute to an individual's overall value for the entity. While water is an extremely familiar good, complications arise concerning the connections, or lack of them, between discernible characteristics (notably taste and smell) and actual properties (contaminant concentrations). In Milan, the unusual taste of tap water and the scale residues left behind by the high carbonate concentration frequently lead to an idea that it is 'filthy and unhealthy'. In addition, exaggerated reports in the media, not least supported by mineral water publicity campaigns, were found to have left many people with the idea that the city's tap water is simply not drinkable. Such factors may cause people to overstate their WTP for groundwater protection since they imagine the current situation to be worse than it really is.

As already described, groundwater represents a stock resource for drinking water, and all the other uses which require certain minimum quality standards to be met, as well as being relevant to various non-use values. The valuation issue, however, is entirely subjective; there is no 'right answer' and therefore the scenario and valuation question, while needing to be informative, should not seek to influence how people think about the resource. Likewise, the offer to change the quantity/quality of its availability must be both politically/institutionally neutral and credible. The handling of this dimension is described in the next section.

The valuation issue is also a challenge for the CVM methodology, since it involves a marginal change in the supply of a complex commodity to which respondents have low 'sensitivity'. In other words, current groundwater quality is on the threshold of being satisfactory in terms of being able to provide safe drinking water (as defined by law) and the policy change for which the benefits need to be quantified involves measures which will prevent a further deterioration in the state of the resource, rather than offer any improvement. The commodity is complex due to the multiplicity and intangibility of its uses. Low sensitivity to the quality dimension is a combined effect of high mineral water consumption levels and the fact that readily discernible taste and smell properties relate to natural features of the aquifer rock substrate rather than pollution phenomena which can be controlled through policy intervention.

A final dimension to considering the valuation question is the respondent's background, in terms of his/her *capability* of making a value judgement of the type required by the study. Independently of difficulties in understanding and prioritising the significance of protecting the resource per se, does the person have market experience especially in respect of the most closely analogous goods, i.e. mineral water prices and water tariffs?

Questions 8, 9, 16 and 23 of the questionnaire (see Appendix 2) of this chapter on whether the person regularly contributes to household income, does grocery shopping and has knowledge of current household outgoings for mineral water and water rates were therefore included to assist in interpreting the reliability of answers given and to better understand the robustness of the respondent's terms of reference.

The scenario presented in the survey is reproduced below:

    I Milan's drinking water is sourced entirely from underground aquifers.

    II Current drinking water supply in the city (only just) complies with standards set by the European Commission for water quality.

    III Production activities, both industrial and agricultural, tend to have a negative impact on groundwater quality which necessitates the introduction of a new, integrated and extensive Water Resource Management Plan relating to Milan.

    IV *Doing nothing*, on the other hand, would lead to a continued rise in the concentration of dangerous substances implicating health risks, worsening of general environmental quality...

    V The implementation of the Management Plan, however, involves additional costs to which Milan residents would have to contribute.

During the pilot stage of the questionnaire, respondents were offered further information to describe key concepts, such as: what groundwater actually is; how water management is currently handled by the Milan authorities; which are the key contaminants of concern and the types of measure implicated in a resource management plan. Cards were prepared for interviewers with information appropriate to ensure that all respondents received a standardised explanation. Since there were no requests for further information in the 50 trial runs, this option was excluded from the main survey and it is assumed that interviewees felt that they knew enough about the valuation issue to answer the WTP questions. This type of serendipity on the part of consumers is also common in real marketplaces where people buy things without knowing exactly how they work (e.g. washing machines) or indeed whether they do (e.g. beauty products).

The valuation question was posed in the form of a *public referendum* in which residents of the city would vote yes or no to establishing a *special fund* for the introduction of a *water resource management plan*. If the majority voted yes, all residents would have to contribute to the fund; it is not a voluntary system (since the water supply infrastructure cannot be adjusted to accommodate differing requirements and the groundwater resource is a classic public good). Regarding effecting the change in provision of the stated environmental good, the tool to be considered was expressed in the form of the development and implementation of a plan to halt further deterioration in groundwater quality. This meant that respondents would have experience of what they would be paying for (in the form of current water supply), rather than requiring them to conceptualise hypothetical improvements which would, in any case, be marginal, since current water quality poses no significant health risks. This formulation of the valuation issue also represents a relatively moderate stance as regards the environmental quality objective, since current groundwater reserves seem to be adequate for the provision of legally compliant drinking water.

Payment vehicle and WTP elicitation method

Payment vehicle and WTP elicitation method are extremely sensitive to the various sources of bias associated with CVM work. While the intention is to be as realistic as possible, it is important to take measures to avoid the possibility that respondents load their answer with other concerns which do not relate to the good/service being valued. The following payment vehicles were considered.

(1) Contribution to a 'Special Fund'.
(2) Increased water tariffs.
(3) A general rise in prices, reflecting the increased costs to producers of further water pollution regulations.

The 'Special Fund' concept allows definition of the payment vehicle to be deliberately kept vague in this survey, owing to local considerations. The Italian water sector is still entirely public and public administrations in Italy are widely considered to be inefficient and unreliable. Indeed, the first scandal which opened the major spate of judicial investigations of public sector corruption broke out in Milan during the late 1980s and continues today. It was therefore decided to avoid direct association between funds raised as a result of the referendum and additional water management responsibilities being handled by the existing water authority due to widespread disillusionment about public sector bureaucratic inefficiency.

The idea of a new system for overseeing water resource management was intended to dissociate the issue from such problems while remaining precise about the financial contribution. A general increase in the price of goods because of the extra costs of pollution control measures to be absorbed by industrial and agricultural producers could have been used to avoid institutional problems but this was considered to be too tangible. To insure ourselves against misjudgement in this respect, however, the WTP elicitation questions included the option to select either increased water tariffs or general prices of consumer goods, for respondents who voted against the Special Fund, and thereby distinguish between respondents who were willing to pay something for water protection and those who perceived no benefits at all.

The WTP elicitation system developed for this survey combined two different elicitation approaches. WTP was first determined via a binary, referendum style question – respondents either accepted or rejected a single 'bid' representing the annual household contribution to the 'Special Fund'. Six different bids, ranging from ITL25 000 to ITL5 000,000, were distributed evenly and randomly within the sample population. Each respondent faced a single bid only. (This is represented by the variable X in Question 2a of the scheme set forth in Figure 6.2.) This was followed by an open-ended WTP question: respondents who had accepted the proposed bid were asked whether they would be willing to pay any more; bid rejecters were asked to suggest a sum that they would be willing to pay between zero and the proposed bid (since they had already declared an intention to support the establishment of the Fund, it was assumed they would be willing to pay something). This elicitation scheme is summarised in Figure 6.2.

While the first WTP elicitation method is considered to be most like a real market situation, where a given item has a price which consumers either accept, and buy the item, or reject, the open-ended WTP question represents an opportunity for respondents to state freely the value they attribute to the environmental resource. This combination of WTP questions is likely to ensure some information on WTP even if the vector of bids would fail in the sense that, for example, a very high proportion of respondents would accept the highest bid. A drawback of the combination, however, is the potential anchoring effect of the initial proposed bid.

A third, and final, WTP figure was elicited near the end of the questionnaire, following on from a series of questions concerning beliefs in the direct substitutability between tap water and mineral water, the wider environmental significance of groundwater resources and general

1. Willingness to support the idea to establish a Special Fund for water resource management (which would involve household contributions to its financing)?

YES →   2a. Accept proposed BID (in the form of a household contribution) ITL *X*?

        YES →   3a. WTP in excess of proposed bid?

               YES →   4. ITL _____

               NO

               DON'T KNOW

        NO →   3b. What WTP between zero and the proposed bid?

               ITL _____. *If nothing*, why?

        DON'T KNOW

NO →   2b. Alternative preference for:

        – higher *water tariffs* → YES / NO

        – increased regulation and operating restrictions on producing activities which would bring about general increases in *consumer goods prices* → YES / NO

        If NO to both, reasons for unwillingness to dedicate more resources to water resource protection?

DON'T KNOW

Fig. 6.2. Willingness-to-pay elicitation scheme.

environmental concerns (Q16–Q20). The aim was to accommodate any awakened appreciation of the benefits of improved water management, without biasing initial responses. The various WTP results are analysed and interpreted in the section on Empirical methods and results.

Questionnaire development

This phase lasted several months, consisting of focus group sessions, open-ended questioning on specific issues and a pilot test of the full questionnaire. The process of questionnaire development and precise definition of the valuation issue are mutually dependent. Only through interaction with the target population were we able to get an idea of key concerns regarding water quality and the information / perceptions base from which these derive.

To obtain meaningful results, the questionnaire had to render the complex set of issues as clearly as possible and elicit answers in an effective and unbiased manner. This calls for questions to be simple, with as much use of the closed-ended form as possible (i.e. multiple choice or yes/no answering). Careful attention was also given to the structure and ordering of questions to ensure that concepts which potentially impact upon the valuation issue are introduced in a logical sequence to minimise the degree to which respondents may be influenced by the questions posed. See Appendix 2 of this chapter for the final version (in English).

An intermediate stage of questionnaire development involved testing reactions to various types of question and alternative forms of information provision. The adopted methodology was fairly detailed face-to-face interviews with individuals (sometimes including family participation) in their homes or approached 'cold' in the street. Widespread interest in the issue of water quality was observed and a significant number of people considered it a major social/environmental issue. Background knowledge was found to be patchy and inaccurate – while most people knew that Milan's tap water supplies came from underground aquifers, understanding of specific pollutants, their origin and the hydrological cycle was virtually non-existent. Only limescale (carbonate deposits) received specific mention and was frequently misinterpreted as a sign of contamination. Some people had strong, but poorly founded and exaggerated ideas about links between water quality and health effects.[7]

Virtually no-one knew what they currently pay for water services. As regards attitudes to groundwater protection, mineral water and the environment in general, a full range of positions was found ranging from pure use value to the idea that groundwater is a crucial resource, in its own right. As regards actual WTP, there was sometimes a missing correlation between a person's willingness to make economic sacrifices and the perceived significance of water protection measures: some who expressed deep concern remained unwilling to pay anything, maintaining that water is a public good, like clean air, to which all people have an equal right and for which they should not be made to pay. The logical weakness of this position, by ignoring the fact that everyone is also at least partly responsible for polluting the resource through their consumption of goods whose manufacture involves the production of effluents, for example, is evident only to experts in environmental valuation. It also reveals the widespread disillusionment in water authorities and perhaps government in general which can lead to strategic answers.

The principal aim of the CVM questionnaire is to supply the information necessary for people to make a reasoned decision *without* biasing their responses. In the case of groundwater protection, it cannot be assumed that non-use types of values will always figure among an individual's perceived benefits. In order to avoid any type of information bias, questions about groundwater protection, the substitutability of tap and mineral water and general environmental concerns were placed after the main valuation questions. Respondents could, however, reconsider their WTP in the light of such considerations with a follow-up modified WTP question.

Another part of questionnaire development was determination of the bid vector, the range of bids to use in the binary choice WTP question. This is crucial to determining eventually the mean WTP and for estimating an aggregate demand curve for groundwater protection, in the form of a special management programme. The pilot test, involving a sample of 50 households, served to refine the bid vector, check any potential problem areas in the questionnaire and also to familiarise the interviewers with the material prior to the full-scale survey.

Execution of the survey

The survey was carried out in July 1994. While the final questionnaire was prepared by the study team, the Dipartimento di Sociologia at the Università di Milano provided professional interviewers, managed the sampling exercise and executed the main survey. Population sampling was via a mix of random sampling (using the telephone book) and a database of individuals who had contacted a particular job agency. The latter assisted in ensuring that targets were heads of household, or at least contributed to family income. The objective was to obtain a minimum of about 200 completed questionnaires, spread across the entire city.

The survey procedure began with interviewers fixing an appointment by telephone and interviews were subsequently carried out face-to-face in the respondent's home. Average interview length was 8.2 minutes. A test for interviewer bias showed no significant influence on WTP estimates.

Empirical methods and results

Sample description

Response rate

The total sample drawn was 244 households. Of this, 197 fully completed questionnaires were obtained. By deliberately excluding those who described themselves as being only temporarily or sporadically resident in Milan (15 individuals) and therefore unlikely to participate in public decision making, the effective response rate rises to 86%. The 30 Milan residents who refused to participate gave reasons, such as time pressures, before being told the subject of the survey. Normally, non-cooperators are assumed to have a zero willingness to pay when calculating total benefits from CVM results (lack of motivation is aligned with lack of concern for the resource being considered) but since interview targets refused to participate without knowing the subject matter, such an assumption is not valid in this study.

The response rate for individual questions was consistently high (over 95%) except for three key areas. Firstly, questions relating to potentially relevant benchmarks in valuing the benefits of water resource protection: expenditure on mineral water (per bottle and per annum, 25% non-response) and actual water tariffs (only 5% of respondents would even hazard a guess). A second difficult area was the open-ended WTP questions. Ten per cent of people who rejected the initial bid in the referendum style section did not provide a concrete sum between zero and the proposed bid which they would be willing to pay, even though they had already voted 'yes' in the referendum, which implies annual payments to the Special Fund. Of those who accepted the bid and said they would pay even more, 28% failed to specify a figure, as was the case with 35% of respondents who wanted to modify their WTP towards the end of the questionnaire. Finally, the question on gross annual household income met with a 14% non-response rate. In Italy, this is considered a relatively low level of refusals for such a delicate issue.

Socioeconomic breakdown

Sex
The sample contains a slightly high proportion of males (55%), probably attributable to the use of the job searchers' database. The actual population of Milan, aged 20 and above (as in the study sample) is 46.5% males.

Age
Age distribution in the sample is representative of the population, for which the average age is 44.75 years. The sample median is the 41–50 age range; 58.6% of the sample is over 40 compared with 55.3% for the population of the city.

Table 6.1. *Sample professional status*

Occupation	No. of cases	% of sample
1. Entrepreneur	3	1.5
2. Freelance	22	11.2
3. Shopkeeper/craftsman	46	23.4
4. Manager/university professor	3	1.5
5. White collar/teacher	50	25.4
6. Skilled worker/technician	1	0.5
7. Unskilled worker	16	8.1
8. Domestic helper	3	1.5
9. Pensioner	18	9.1
10. Unemployed	5	2.6
11. Housewife	12	6.1
12. Student	18	9.1
Total	197	100

Education

Everyone in the sample has at least some schooling, the majority holding a high-school diploma or equivalent. University graduates constituted 14.4% of the sample, which is higher than the general figure, again due to the job agency database.

Professional status

The largest shares of the sample consist of shopkeepers/craftsmen and white collar workers. The full breakdown is shown in Table 6.1.

Household typology

Average family size is 2.9; the majority of households number three to four people. Only three respondents lived in households of over five people. The most common household structure is a couple plus one child (48.8% of sample). Over 97% of respondents live in an apartment building (as opposed to single or two family houses) and 61% were home owners, the remainder being tenants. This information helps to attribute the widespread low awareness of current water rates (see sensitivity and calibration of responses, below).

Water consumption behaviour and attitudes

Only 27% of respondents claimed to habitually drink tap water, as opposed to mineral water, although over 99% use tap water for cooking, making tea and coffee, etc. Eighty-nine per cent of respondents said they regularly consumed mineral water, this rising to 91% when they were questioned about household (family) behaviour. Asked specifically about their opinion of Milan's municipal water supply, over half (54%) declared it satisfactory or better. Follow-up questions about tap water quality revealed concern about chemical pollution (58%), health risks (52%) and poor taste (36%). Interestingly, these figures imply that even though some respondents had specific quality concerns they still considered the municipal supply to be satisfactory.

Specific questions about motives for mineral water consumption revealed tap water substitution as by far the most significant factor (73%), followed by 'habit' (42%), e.g. someone else in the household buys it anyway, and particular organoleptic properties (taste, bubbles etc.). Secondary motives were specific therapeutic features or health reasons, including the choice of mineral water instead of wine or other soft drinks. Thirteen per cent of households had installed a domestic water filter of some type (including systems which only reduce water hardness).

Sensitivity and calibration of responses

WTP answers need to be interpreted in the context of a very wide set of factors that influence people's decision-making. First, because of the hypothetical nature of the scenario to be valued and payment elicitation mechanism, questions which reveal (at least subjectively) a person's 'sensitivity' to the central valuation issue can be useful. Secondly, WTP figures should be compared to actual purchasing behaviour for related goods and services to see, at least, whether the orders of magnitude of WTP estimates are reasonable. While it is impossible to address all potentially relevant issues, not least because these vary widely from person to person, a number are considered below.

Familiarity with market decisions
The sampling strategy (partly using a job agency database) helped to ensure that a large proportion of respondents were generally familiar with

having to make purchasing decisions and therefore more likely to provide a responsible, reasoned answer to the WTP questions. Two questions concerning contribution to household income and grocery shopping were inserted into the questionnaire to check for this. It was found that 81% and 61% respectively of respondents regularly carried out these activities.

Awareness of current water tariffs
WTP for an incremental change in the quality of water services would be expected to relate to current water charges (even if, in practice, these are often set outside the context of perfect market conditions and relate to specific political considerations; in the consumer's mind this is not necessarily an issue). In Milan, however, few of the city's residents are aware of what they pay for water services. Tariffs are included in a bundle of charges owners pay to the central administration of apartment buidings (along with general maintenance, waste services etc.). For people in rented accommodation, the situation is even more opaque as such costs are normally paid directly by the landlord and included in the rent package. Furthermore, Italian water tariffs are among the lowest in Europe and the province of Milan has tariff levels among the lowest in the country so people are only very slightly sensitive to these payments, and this disappears in the company of more significant utility items such as electricity, gas and waste collection services. Indeed, only 4.6% (9 respondents) claimed to know what they currently pay for water. Of these, the estimated figures were rather diverse (between ITL60 000 and 400 000) and the mean (ITL171 110 per annum) is higher compared to the actual average Milan water bill. According to information supplied by the competent body the average household actually pays ITL60 000–100 000 per annum.

General environmental concerns
One of the final sections of the questionnaire contained some general attitude questions, designed to give a broad idea of the respondent's level of environmental concern, appreciation of environmental protection issues, and the position of water quality among these. One question asked whether groundwater was a resource to be protected, as a natural resource, irrespective of its direct use as a source of drinking water. The overwhelming majority (96.5%) answered 'yes'. A second question asked the respondent to state his/her two main environmental concerns, selected from a comprehensive list of six issues. The findings are summarised in Table 6.2.

Table 6.2. *Main environmental concerns*

Concerns	% 'yes'
1. Air pollution	66.5
2. General deterioration in environmental quality	59.0
3. Waste disposal	25.6
4. Urban traffic and noise	21.6
5. Global issues	15.9
6. Drinking-water quality	11.3

'General environmental degradation' (meaning large-scale and diffuse phenomena such as marine and river pollution, aquifer contamination, disrupted landscapes and urbanisation) received the largest number of votes (59%) after air pollution (66.5%). Poor drinking-water quality happened to obtain the fewest, although it can be argued that this is a knock-on effect of a general deterioration in environmental quality – at least to some extent.

Consistency tests and correlations

Certain relationships were explored to assess sample representativity. Positive correlations were found between per capita income (i.e. total household income divided by number of occupants) and (1) education or (2) mineral water expenditure (also on a per capita basis), which was to be expected. The relationships between whether respondents drink tap water, their opinion of tap water and mineral water consumption were studied in detail, in order to better understand the motivations underlying stated WTP.

On average, people who drink tap water had a more positive opinion of tap water, compared with non-drinkers of tap water. The difference was large enough to cause a rejection of a null hypothesis of independence between opinion and drinking habit(χ^2(1 df) = 33.24). In line with this, average mineral water expenditure in the subsample which habitually drinks tap water was 44% less than those who said they did not drink tap water. Independence, this time between mineral water expenditure and drinking habit, could again be rejected (χ^2(4) = 48.04). This reinforces the indication that tap-water drinkers have a better opinion of Milan's drinking water supply; if this were not the case, they

Table 6.3. *Sample breakdown according to use of tap water for drinking*

Drink tap water:	Yes	No
Mean income (ITL 1000/person per yr)	15 700	19 800
Mean MW consumption (ITL 1000/person per yr)	80.9	144.6
Proportion regarding tap water as 'excellent', 'good' or 'satisfactory' (not 'bad' or 'terrible')	0.885	0.420

Note:
MW, mineral water.

Table 6.4. *Sample analysis according to use of mineral water and/or tap water for drinking*

Drink mineral water	Drink tap water	No. of individuals	% of sample
Yes	Yes	30	15.2
Yes	No	140	71.1
No	Yes	23	11.7
No	No	4	2.0
Total		197	100

would simply buy mineral water instead. The estimates in Table 6.3 also suggest that people with relatively low income tend to drink tap water. However, a null hypothesis of independence between income and drinking habit could not be rejected (($\chi^2(4) = 1.78$). This indicates that, for example, a low income does not necessarily imply a choice of tap water as drinking water.

Moreover, opinion of tap water correlates positively with a dummy variable representing whether the respondent drinks tap water, and negatively with a dummy for mineral water consumption. In other words, people who do not drink tap water and/or drink mineral water tend to have a lower opinion of tap water. Table 6.4 shows how the two subpopulations overlap (tap-water versus mineral-water drinkers). Non-drinkers of tap water constitute the majority (73.1%); 15.2% of the sample uses both sources of drinking water.

Splitting the sample again between people who drink tap water and those who do not, 47% of tap water drinkers stated that they considered tap water and mineral water to be directly substitutable whereas only 26% of non-tap-water drinkers would agree. While the intention behind the question was

Table 6.5. *Reactions to payment vehicle questions*

Reactions	No. of individuals	% of sample
1. Accepted a Fund	144	73.1
2. Accepted increased water tariffs only	3	1.5
3. Accepted regulations only	23	11.7
4. Accepted both increased tariffs and regulations	4	2.0
5. Accepted none of the payment vehicles	18	9.2
6. Don't know	5	2.5
Total	197	100

to see whether respondents recognised the wider, non-private (externality) advantages of good quality tap-water supply, and one would therefore expect tap-water drinkers to be more sensitive to the non-substitutability, the answers to this question suggest a different reasoning based more on the distinctive features of mineral water.

Benefit estimates

Reactions to the payment vehicle

The block of WTP questions was opened by a question on whether the respondents in the case of a referendum would be in favour of the payment vehicle selected for the WTP questions, i.e. contributions from consumers to a Special Fund for Resource Management. This payment vehicle turned out to be quite successful, since about three out of four respondents answered that they would vote in favour of such a fund. Those who opposed the establishment of a fund met questions on whether they would instead be in favour of increased water tariffs or an introduction of restrictions and regulations for industry and agriculture. When the latter payment vehicle was suggested, it was emphasised that such restrictions and regulations would involve an increase in consumer prices. Table 6.5 summarises the answers to all payment vehicle questions. They are also illustrated by Figure 6.3.

The binary choice WTP question

As was explained earlier, respondents who accepted the payment vehicle of contributions to a Special Fund then met a question on whether they

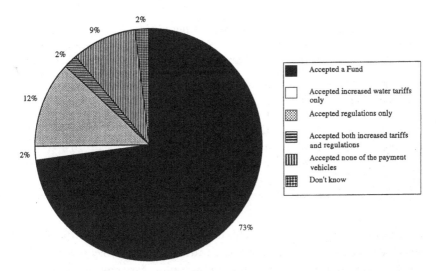

Fig. 6.3. Reactions to payment vehicle questions.

would be willing to pay a specific annual household contribution. Six different amounts of money or 'bids' ITL25 000, 100 000, 500 000, 1 000 000, 2 500 000 and 5 000 000 were randomly distributed among the sampled households. Since mean WTP estimates are sought, an important criterion for the design of the bid vector was that basic information on the tails of the distribution of 'yes' answers should be obtained. An example of such a basic piece of information is for which bid there is no acceptance at all. The bid vector that was employed turned out to work rather well in this respect: as Table 6.6 shows, there were no 'yes' answers to the highest bid. The lowest bid did not, however, result in a 100% acceptance. Table 6.6 also shows that the number of respondents who gave a 'don't know' answer was low (4.2%) and no respondent refused to answer this WTP question.

The bids and the proportion of respondents who accepted each bid can be used for estimating an aggregate demand curve for a discrete commodity, i.e. to introduce the Water Resource Management Plan or not. Mean WTP can then be computed as the area under this curve (see, for example, Johansson, 1993, pp. 62–4, and Söderqvist, 1995). That is:

$$E[\text{WTP}] = \int_0^\infty D(a)\,da \qquad (6.5)$$

Table 6.6. *Responses (Willingness to pay) to offered bids*

Bid (ITL)	Yes	No	Don't know	Sum	% Yes	% Yes (Ayer)
25 000	17	3	0	20	85.0	86.7
100 000	22	0	3	25	88.0	86.7
500 000	9	16	1	26	34.6	34.6
1 000 000	6	17	1	24	25.0	25.0
2 500 000	5	19	0	24	20.8	20.8
5 000 000	0	24	1	25	0.0	0.0
Total	59	79	6	144		

where a is the 'price' of the discrete commodity – the bids constitute some possible prices – and $D(a)$ is the aggregate demand curve. In equation (6.5), it is assumed that no respondent requires a monetary compensation for accepting an introduction of the Plan, i.e. that WTP is non-negative.

One possible way of estimating the curve is to follow the non-parametric method of Ayer *et al.* (1955) (see, for example, Kriström, 1990, and Söderqvist, 1995, for applications). This method involves a conversion of the proportion of 'yes' answers to a monotone non-increasing sequence. The data in Table 6.6 imply the sequence in the table's seventh column. Let us first assume that the proportion of 'yes' answers would be equal to unity for a zero bid, that the proportion would be equal to zero for bids greater than ITL5 000 000 and that the proportions in between the used bids (including zero) can be approximated by a linear interpolation. The data available and these assumptions result in the aggregate demand curve shown by the continuous line in Figure 6.4. This curve implies an annual mean household WTP of ITL1 083 000. Median WTP is estimated as the amount of money that, according to the curve, corresponds to the 0.5 proportion, i.e. about ITL350 000.

At least one of the assumptions underlying this calculation of mean WTP is, however, partly unsatisfactory. It was assumed that the proportions in between ITL2 500 000 (0.208) and 5 000 000 (0) can be approximated by a linear interpolation. As can be seen from Figure 6.4, this interpolation results in a triangle that constitutes almost 25% of the total area under the aggregate demand curve. In reality, the curve may approach the vertical axis much quicker than in this linear interpolation, thus reducing mean WTP substantially. A helpful piece of information

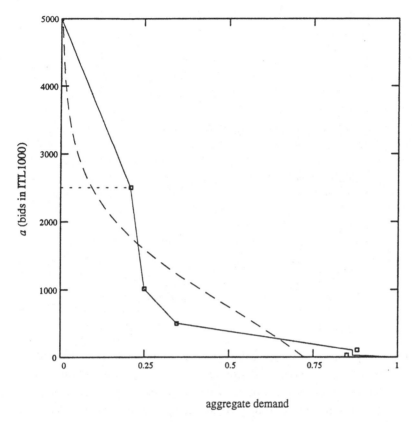

aggregate demand

Fig. 6.4. Aggregate demand curve. Boxes, observations; continuous line, non-parametric specification; dashed line, logit specification; dotted line, truncation at ITL2 500 000.

would be the lowest bid above ITL2 500 000 for which the proportion of 'yes' answers is zero.

An indication of the size of this bid can be obtained from the open-ended WTP questions that followed the binary choice WTP question. In these questions, the respondents who accepted the bid they met were asked whether they would be willing to pay more than the bid, and the respondents who rejected a bid were asked how much they would be willing to pay between zero and the bid they rejected. The answers to these questions are analysed in detail below. For the present purposes, suffice it to note that no respondent stated a higher maximum WTP than ITL2 500 000. To be precise, the five respondents who accepted the ITL2 500 000 bid were not

Table 6.7. *Ml estimation results for the logit model*
$\text{BIDYES}_i = 1 / [1 + \exp(- (\alpha - \beta \text{BIDLIRE}_i))]$, $i = 1, \ldots, 144$

Coefficient	Estimate	SE	t value [p]
α	0.95619	0.2727	3.507 [0.00]
β	0.0012956	0.0002793	4.639 [0.00]

Log-likelihood = -69.36, pseudo-$R^2 = 0.29$,
$\chi^2(1) = 56.18$, significance level of $\chi^2 = 0.000$

Actual and predicted outcome
(frequency):

Actual	Predicted		Total
	0	1	
0	62	23	85
1	11	48	59
Total	73	71	144

Note:
SE, standard error.

willing to pay anything more, none of the respondents who rejected the
ITL5 000 000 bid stated a WTP greater than 1 500 000 bids and the highest
maximum WTP of the respondents who accepted any of the 72 500 000 was
2 000 000. This information suggests that the proportion of 'yes' answers is
zero for any bids greater than ITL2 500 000. The dotted horizontal line in
Figure 6.4 illustrates this adjustment of the aggregate demand curve. The
mean household WTP is now ITL823 000, and median WTP is unchanged,
i.e. about ITL360 000.

Another way of estimating mean (and median) WTP is to use some para-
metric method. This involves an assumption that the distribution of 'yes'
answers follows a specific probability model. The most commonly
employed model in CVM studies is the logit model. The results of the
estimation of a simple logit model are found in Table 6.7. Individual data
were used for the estimation, and the dummy variable BIDYES takes the
value of unity in the case of acceptance of a bid, and zero otherwise. The
explanatory variable BIDLIRE is simply the bids in thousands of ITL. The
estimation was done by the LOGIT command of Limdep 6.0, which
implied the use of the maximum likelihood (ML) method (see Greene, 1991,
p. 484). It is evident from the table that the coefficient of BIDLIRE is

significantly different from zero and has an expected sign: the probability of a 'yes' answer decreases with the size of the bid. The predictive power of the estimated model is rather good: about 76% of the predicted outcomes are equal to the actual ones.

Mean WTP can now be estimated, given different truncations of the logistic distribution. The basic assumption of equation (6.5) implies that:

$$E[\text{WTP}] = \int_0^\infty D(a)\mathrm{d}a = \int_0^\infty (1/\{1 + \exp[-(\alpha-\beta a)]\})\mathrm{d}a$$
$$= (1/\beta)\{\ln[1 + \exp(\alpha)]\} = \text{ITL}989\,000$$

The discussion above when the non-parametric method was followed suggests, however, that the distribution should also be truncated from above at ITL 2 500 000, i.e.:

$$E[\text{WTP}] = \int_0^{2500} D(a)\mathrm{d}a = \int_0^{2500} (1/\{1 + \exp[-(\alpha-\beta a)]\})\mathrm{d}a$$
$$= (1/-\beta)\{\ln[\exp(\alpha-2500\beta) + 1] - \ln(\exp(\alpha) + 1]\}$$
$$= \text{ITL}914\,000$$

The approximate standard errors of these estimates are ITL161 000 and 117 000, respectively, which implies that mean WTP in both cases is significantly different from zero at conventional levels of significance.[8] The estimator for median WTP is α/β, which results in an estimate that is not influenced by the truncations, ITL738 000. The estimated aggregate demand curve based on the logit model is shown by the dashed curve in Figure 6.4. The left part of the dotted horizontal line indicates the truncation at ITL2 500 000.

Open-ended WTP questions

As mentioned above, the respondents who accepted the bid they met in the binary choice question were subsequently asked whether they would be willing to pay more than the bid, and the respondents who rejected the bid were asked how much they would be willing to pay between zero and the bid they rejected. This means that each respondent had the chance of stating a specific maximum WTP amount (variable: OPENWTP). For the respondents who accepted a bid, but were not willing to pay any extra amount above the bid, the bid was regarded as the value of OPENWTP for these respondents.

Table 6.8. *Statistical description of* OPENWTP *and* MODIFWTP

	Mean (ITL)	Median (ITL)	Range (ITL)	SD (ITL)	No. of observations
OPENWTP	417 040	200 000	0–2 500 000	541 360	130
MODIFWTP	447 800	200 000	0–2 500 000	565 860	132

Note:
SD, standard deviation.

All respondents were also given an opportunity to modify their WTP towards the end of the interview. This opportunity to state a modified maximum WTP (variable: MODIFWTP) was given after some behaviour and attitude questions on, *inter alia*, the importance of protecting ground-water in relation to environmental protection in general. Also the respondents who did not accept contributions to a Special Fund as a payment vehicle met this question (although only two of them stated a WTP).

Table 6.8 reports summary statistics for OPENWTP and MODIFWTP. The mean WTP estimates are ITL417 000 and 448 000, respectively. The mean WTP estimate based on MODIFWTP is thus only about 7% greater than that based on OPENWTP. In fact, only 13 respondents modified their value of OPENWTP. All of them chose to increase their maximum WTP. The largest difference between MODIFWTP and OPENWTP was for a respondent who doubled his maximum WTP from ITL1 000 000 to 2 000 000.

Comparison of mean WTP estimates: compliance bias and anchoring effects

Table 6.9 summarises the mean (and median) WTP estimates computed so far. The approximately two-fold difference between mean WTP estimates based on the answers to the open-ended questions and those based on data from the binary choice question calls for a closer examination. In the first year, there may have been a certain degree of compliance bias, also known as 'yea saying'. This phenomenon is frequently referred to in critiques of referendum style CVM studies and the allegation is that individuals accept the bid in order to free themselves as quickly as possible from the interview, because it is easier than thinking seriously about the question, given the artificiality of the survey context, etc.

Table 6.9. *Mean and median* WTP *estimates in* ITL

Model	Mean WTP	SD of mean WTP	Median WTP
Binary choice: non-parametric	1 083 000		360 000
Binary choice: non-parametric (truncated at ITL 2 500 000)	823 000		360 000
Binary choice: logit model	989 000	161 000	738 000
Binary choice: logit model (truncated at ITL 2 500 000)	914 000	117 000	738 000
Open-ended WTP: OPENWTP	417 000	37 500	200 000
Open-ended WTP: MODIFWTP	448 000	39 500	200 000

Considering the data in more detail, Table 6.10 reports what values of OPENWTP followed after an acceptance or a rejection of a certain bid. This table reveals the observation made earlier than none of the respondents who accepted a bid was willing to pay more than ITL2 500 000 and none of the respondents who rejected the ITL5 000 000 bid stated a WTP above ITL1 500 000. Again, this piece of information provided by OPENWTP suggests that the mean WTP estimates based on the untruncated binary choice models are exaggerated. However, there are also problems associated with OPENWTP and MODIFWTP.

More precisely, Table 6.10 indicates that anchoring effects are present in the sense that the values of OPENWTP are not independent of the bids that were randomly distributed among the respondents. Table 6.10 reports that for the respondents who met the lowest bid (ITL25 000), the mean value of OPENWTP is 65 750, which is considerably lower than the overall mean (417 000). For the high bids, the mean values of OPENWTP are higher than the overall mean. For example, the mean value of OPENWTP for the respondents who met the 2 500 000 bid is 895 650. The same pattern is present for MODIFWTP. These large differences in mean values of OPENWTP and MODIFWTP arise despite the fact that all respondents (irrespective of what bid they met) could freely state a maximum WTP.

The presence of an anchoring effect may also be revealed by a simple test where OPENWTP (or MODIFWTP) is regressed on BIDLIRE (cf. Mitchell and Carson, 1989, p. 241). Such a regression reveals that the coefficient estimate of BIDLIRE is positive and significantly different from zero. Also an

Table 6.10. *BIDLIRE and OPENWTP: the number of respondents who stated a certain maximum wtp after the binary choice*

BIDLIRE (ITL)	OPENWTP (ITL)	
	'Yes' to bid	'No' to bid
25 000	25 000: 6, 45 000: 1, 50 00: 3, 70 000: 1, 100 000: 4, 200 000: 1 and 300 000: 1	0:3 (Note: These zero WTP answers may alternatively be intrepreted as a late protest against the Fund)
	Example: '300 000: 1' above means that among those who accepted the bid of ITL25 000, one respondent stated ITL300 000 as his maximum WTP. Mean: 65 750	
100 000	10 000: 16, 120 000: 1, 150 000: 1 and 200 000: 4	(No 'no' responses)
	Mean: 121 360	
500 000	500 000: 8 and 700 000: 1	50 000: 2, 10 000:2, 150 000:2, 200 000: 4 and 250 000:1
	Mean: 317 500	
1 000 000	1 000 000: 4 and 1 500 000: 2	100 000: 7, 200 000: 1, 250 000: 1, 300 000: 5, 350 000: 1, 450 000: 1 and 500 000: 1
	Mean: 476 090	
2 500 000	2 500 000: 5	0: 1, 100 000:1, 200 000: 3, 300 000: 2, 400 000: 1, 500 000: 6, 600 000: 2, 1 000 000: 1 and 1 200 000: 1
	Mean: 895 650	
5 000 000	(No 'yes' responses)	5 000: 1, 25 000: 1, 100 000: 4, 200 000: 3, 500 000:3, 600 000: 1, 700 000: 1, 1 000 000: 7 and 1 500 000: 1
	Mean: 560 450	

alternative specification, where BIDLIRE is transformed to five dummy variables (there were six bids), results in positive and significant coefficient estimates for most of the dummy variables (see Tables 6.11A and 6.11B or ordinary least squares (OLS) regression results). The presence of this anchoring effect suggests that the bids themselves were often important

Table 6.11A. *OLS* *estimation results for the* *anchoring model*

$OPENWTP_i = \alpha + \beta BIDLIRE_i + \epsilon_i, i = 1, \ldots, 130$

Coefficient	Estimate	SE	t value [p]
α	251.47	59.75	4.209 [0.00]
β	0.10593	0.02535	4.178 [0.00]
Adjusted $R^2 = 0.11$			

Note:
SE, standard error.

Table 6.11B. *OLS* *estimation results for the* *anchoring model*

$OPENWTP_i = \gamma_0 + \gamma_1 BID5000_i + \gamma_2 BID2500_i + \gamma_3 BID1000_i + \gamma_4 BID500_i + \gamma_5 BID100_i + \epsilon_1, i = 1, \ldots 130$

Coefficient	Estimate	SE	t value [p]
γ_0	65.750	105.1	0.625 [0.53]
γ_1	494.70	145.3	3.405 [0.00]
γ_2	829.90	143.8	5.773 [0.00]
γ_3	410.34	143.8	2.854 [0.01]
γ_4	251.75	148.7	1.693 [0.09]
γ_5	55.614	145.3	0.383 [0.70]
Adjusted $R^2 = 0.25$. $F(5, 124) = 9.40$			
Significance level of $F = 0.000$			

Note:
SE, standard error.

influences on what WTP the respondents stated. This may indicate some uncertainty among people about their 'true' preferences for groundwater protection.

The anchoring effects make it difficult to interpret the answers to the open-ended WTP questions. As a consequence, we will, in the following, focus on the mean WTP estimates of the binary choice WTP questions. The anchoring effects also throw doubt on the procedure used above to truncate the binary choice models at ITL2 500 000, since answers that may have been influenced by the bids are used for the truncation. However, we will stick to this procedure, since none of the respondents stated a WTP greater than ITL2 500 000 despite the fact that subjects who met bids such as ITL5 000 000, 2 500 000 and 1 000 000 seem to have been induced to state relatively high WTP amounts in the open-ended questions. The mean WTP estimates of ITL823 000 and 914 000 (on average, 868 500) will thus be preferred in the following. A sensitivity analysis of these estimates is given in the next section.

As to the potential compliance bias, there is a clear tendency among the respondents who accepted a bid to *not* state a higher WTP than the bid in the subsequent open-ended WTP question. This tendency may indicate a certain degree of 'yea saying'. While we consider this indication to be important in an evaluation of the resulting benefit estimates, it should be acknowledged that there are alternative interpretations of this tendency. For example, it could be due to preference uncertainty.

Sensitivity analysis, sample and aggregate WTP estimates

Trimmed mean WTP

It is common in CVM studies to analyse the influence on WTP of extreme opinions among respondents. Such an opinion may be defined as a case when a person is willing to pay an unusually large proportion of his or her income for the good in question, in this case the establishment of the Water Resource Management Plan. One way of computing trimmed mean WTP estimates is to exclude respondents with such opinions from the analysis. Such an exclusion must not be made routinely, since there is in general no reason to regard unusual opinions as observations not worth paying any attention to. However, it can be of interest to study the sensitivity of mean WTP estimates to such opinions.

Table 6.12 presents summary statistics for the variable BIDINC. This variable is simply accepted bids in the binary choice WTP questions divided by stated income. A non-response rate of 14% for the question on income explains why the number of observations for BIDINC is somewhat lower than the total number of 'yes' answers to the bids. The table shows that the maximum value of BIDINC is about 6%. 'Unusual' observations can be defined as the outside values of the distribution of BIDINC. Following the conventional definition of such values, the outside values are those for which BIDINC > 0.0437.[9] It turns out that two

Table 6.12. *Statistical description of* BIDINC *(accepted bid divided by annual income)*[a]

	Mean	Median	Range	SD	No. of observations
BIDINC	0.0099	0.0025	0.00023–0.0617	0.0143	52

Notes:
SD, standard deviation.
[a] The close-ended income question in the questionnaire gave information on according to the following income groups that households belong to: <10, 11–20, 21–30, 31–50, 51–70, 71–100, >100 (ITL million). In order to compute BIDINC, middle values of the groups were used, i.e. 5, 15.5, 25.5, 40.5, 60.5, and 85.5. For the group of >100, the value was arbitrarily set to 110.5.

Table 6.13. *Untrimmed and trimmed mean* WTP *estimages in ITL*

Type of exclusion of respondents	Mean annual household WTP	No. of observations
Non-parametric, truncated at ITL2 500 000, all respondents	823 000	144
Logit model, truncated at ITL2 500 000, all respondents	914 000	144
Non-parametric, truncated at ITL2 500 000, outside values of BINDIC excluded	769 000	142
Logit model, truncated at ITL2 500 000, outside values of BINDIC excluded	801 000	142

observations meet the criteria for outside values. These two respondents both accepted a 2 500 000 bid and reported an income between ITL31 and 50 million.

The trimmed mean WTP estimates of Table 6.13 are obtained when these two observations are excluded from the analysis. The trimmed estimates of the non-parametric model and the logit model differ less than the untrimmed ones. When compared with the average untrimmed mean WTP estimate of the two models, i.e. ITL868 500, the average trimmed mean WTP estimate of the two models (ITL785 000) is about 10% lower. A similar reduction follows for a trimming of the mean WTP estimates based on the open-ended WTP question.

Protests, non-responses, and sample and aggregate WTP estimates

The mean WTP estimates presented hitherto are valid only for the respondents who answered WTP questions in the interviews. However, there are also respondents who rejected the Fund payment vehicle and said that they would accept another payment vehicle, but none the less did not give any answer to the final WTP question. There were also respondents who rejected all suggested payment vehicles, and there were persons in the sample who refused to participate in the survey altogether. It is important to study the sensitivity of the mean WTP for assumptions regarding the WTP of such groups. Some assumptions that result in a conservative mean WTP estimate for the total sample will be used here.

The total sample size was 244 households. They can be divided into the following groups:

A 144 observations are available for the binary choice WTP question, and the average mean WTP estimate of the two binary choice models, ITL868 500, is valid for this group of respondents.

B 7 interviewees did not accept the idea of the Fund, but said that they would accept increased water tariffs. The mean WTP estimate of ITL868 500 can be assumed to be valid for this group too, since the acceptance of increased water tariffs reveals that a household willingness to pay exists.

C 23 persons did not accept the idea of the Fund or increased tariffs, but said that they supported the introduction of restrictions and regulations for industry and agriculture. It was emphasized in the interviews that such an introduction would cause an increase in consumer prices. However, the support of this payment vehicle indicated a desire that someone else should pay, and a zero WTP is therefore assumed to be valid for all persons in this group, except for one respondent who, towards the end of the interview, said that he was willing to pay ITL100 000.[10]

D 18 persons did not state any WTP, and did not accept any of the proposed payment vehicles. A zero WTP is assumed also for this group.

E 5 persons did not know whether to support the establishment of the Fund or not. A conservative assumption is to assign a zero WTP to this group, except for one person who eventually stated a WTP of ITL150 000 (cf. note 10).

F 47 persons were non-respondents. A common conservative assumption in CV studies is to assign a zero WTP to all non-respondents. However, as was mentioned earlier, this assumption is hardly valid for 30 of the non-respondents, since they refused to be interviewed before they were told the subject of the study. A zero WTP will consequently be assumed to be valid for only 17 of the non-respondents. The mean WTP estimate will be assumed for the remaining 30 persons.

All these assumptions result in the following *sample estimate of mean annual household WTP*:

$$868\,500 \times (181/244) + 100\,000 \cdot (1/244) + 150\,000 \times (1/244) + 0 \times (61/244) \approx ITL645\,000$$

Since the number of households living in Milan is approximately 533 000, this implies an aggregate WTP of ITL343 785 million. If the conventional assumption of a zero WTP for *all* non-respondents were to be adopted, the mean estimate would decrease to ITL538 000, and the aggregate figure to ITL286 754 million. These aggregate figures should not be confused with the present value of annual payments to the Fund. The size of the present value depends on what discount factor and time horizon are regarded as reasonable.

Estimation of a WTP function

It is a common practice in CVM studies to estimate a WTP function, i.e. a function that relates WTP to variables that are supposed to have an influence on the stated WTP. Such an estimation should be seen largely as explorative, but sometimes economic theory indicates what type of an influence a certain explanatory variable could be expected to have. If this prediction does not turn out to be supported by the estimated WTP function, there are reasons to have a closer look at what circumstances could have caused the unexpected results.

Given that a WTP function can be estimated successfully, there are interesting ways to use this function. For example, it can be assumed that the estimated function is valid not only for the population from which the sample used is drawn, but also for other populations. These populations can then be characterised by certain values of the explanatory variables that in turn produce a certain WTP value when inserted into the estimated function. An estimated WTP function may also be of some help when non-respondents are taken into account. Given that it is known what values the explanatory variables take for a non-respondent or a group of non-respondents, the estimated function predicts the WTP of these individuals. Such a use of the WTP function presupposes that the estimated WTP function is valid both for the group of respondents and for the group of non-respondents.

Selection of explanatory variables

What explanatory variables should be included in this study's WTP function? Besides income (INCLIRE), it could be expected that people's consumption of substitute (or complement) private goods influences their WTP. In the case of protection of Milan's groundwater, there are two substitute private goods to consider: people have the option of buying mineral water and they can install domestic water filters. As was explained earlier, these substitutes are not perfect. Recall that the Water Resource Management Plan involves not only the protection of drinking-water quality, but also that of an environmental resource (groundwater). Moreover, bottled mineral water has additional characteristics when compared with tap water, as regards distribution and taste, therapeutic properties, etc., and domestic water filters are not always efficient in removing contaminants.

However, there may also be people who regard mineral water or tap water passed through by domestic filters as perfect substitutes for tap water from protected drinking-water sources. Their expenditures on mineral water or water filters can be viewed as their defensive expenditures to prevent or to counteract the consequences of non-protected water resources. Such defensive expenditures can be used as a measure of the WTP for (a marginal increase in) protection of drinking water sources (see Freeman, 1993, pp. 114–15). This suggests that, for some people, there may exist a positive relationship between mineral water expenditure and WTP. For water filter expenditure, the matter is somewhat less clear-cut. A household which has purchased a filter may regard it as a long-term investment, which means that it would not be very interested in paying extra for a water resource protection plan.

In order to tackle these issues, a dummy variable for 'water filter installed'

(FILT) was included in the WTP function to be estimated. As regards mineral water expenditures, the basic idea is that some people buy mineral water as a result of dissatisfaction with tap-water quality. It is such defensive expenditures that should be included as an argument in the WTP function, *not* mineral water expenditures that arise from, for example, consuming mineral water as an alternative to alcoholic drinks. This means that a variable defined as annual household mineral-water expenditure per household member (PMYEAR) is likely to be too crude. A more refined variable (PMWCA) was defined in the following ways:

PMWCA = 0 for respondents who were not mineral-water consumers or who consumed mineral water because of 'substitute for other types of drinks (alcoholic or non-alcoholic)' or 'habit'.

PMWCA = 0.5 × PMYEAR for respondents who were mineral-water consumers and stated 'substitute for other types of drinks' or 'habit' as only one of two main reasons for their consumption.

PMWCA = 1 × PMYEAR for respondents who were mineral-water consumers and did *not* state 'substitute for other types of drinks' or 'habit' as a main reason for their consumption.

PMWCA gave some, but not ideal, information on defensive expenditures, since the other reasons the respondents gave for their mineral-water consumption were far from unambiguous. For example, 'aesthetic properties' could refer to people's preference for mineral water because of tap water's bad taste, which they (incorrectly) think is caused by pollution, but it could also refer to a preference for bubbles (fizziness) in drinking water, i.e. something that cannot be expected from even the purest tap water. In the WTP function to be estimated, both PMWCA and the crude PMYEAR were tried as explanatory variables.

INCLIRE, FILT, PMWCA and PMYEAR are described statistically in Table 6.14. This table also includes a description of a set of other explanatory variables which concern household characteristics and attitudes. These are conventional socioeconomic characteristics such as sex, age and eduction, but also two dummy variables (SENS1 and SENS2) that deserve special attention. It was suggested on pp. 134–7 that respondents who are not regular contributors to household income (SENS1 = 1) and/or do not regularly do the grocery shopping (SENS2 = 1) may have a weaker ability to make robust value judgements. However, it is unclear how this weaker ability influences the stated WTP.

Table 6.14. *Statistical description of the variables of the WTP function*

Variable	Mean	Median	Range	SD
BIDYES: Dummy variable for acceptance of the bid given in the binary choice WTP question (dependent variable)	0.40566	0	0/1	0.49335
BIDLIRE: Bid in 1000 ITL given in the binary choice WTP question	1574.5	1000	25–5000	1769.3
SEX: Dummy variable for female respondent	0.47170	0	0/1	0.50157
AGEYEAR: age of respondent in years	42.792	45.5	18–65.6	13.609
EDUC: Education of respondent in points (1 = no education, 2 = primary school, 3 = school leaver certificate, 4 = high school diploma, 5 = university degree)	3.8585	4	2–5	0.77384
CHILD: Dummy variabler for at least one child in the household	0.68868	1	0/1	0.46523
SENS1: Dummy for that the respondent is *not* a regular contributor to household income	0.15094	0	0/1	0.34969
SENS2: Dummy for that the respondent does *not* regularly do the grocery shopping	0.34906	0	0/1	0.47894
PINCOM: Annual gross household income in 1000ITL divided by the household size	19183	15125	1667–110500	16616
TWOPI: Opinion of Milan tap water (1 = excellent, 2 = good, 3 = satisfactory, 4 = bad, 5 = terrible)	3.4340	3	1–5	1.0330
FILT: Dummy for possession of domestic water filter	0.15094	0	0/1	0.35969
PMYEAR: Annual household mineral water expenditures in 1000 ITL divided by household size	125.45	104.00	0–728	110.17
PMWCA: Proxy for annual defensive household mineral water expenditures in 1000 ITL divided by household size (see text for discussion of this variable)	95.407	69.333	0–589.3	94.265

Notes:
Number of observtions: 106, i.e. 106 respondents gave data for all variable above.
SD, standard deviation.

The WTP function is a multivariable extension of the logit model estimated on pp. 148–53. The results of an ML estimation of this WTP function are presented in Table 6.15.

Findings from the WTP function

Table 6.15 shows that the model predicts the actual outcomes in about 76% of all cases. This is a respectable figure, but it is not higher than that for the two-variable model of Table 6.7. Moreover, among the explanatory variables, only the coefficient of BIDLIRE is significantly different from zero at a low level of significance. Also CHILD seems to be of importance. The positive sign of its coefficient estimate implies that households with at least one child have a higher WTP than other households. An encouraging finding is that the coefficient estimates of PINCOM (annual income per person) and PMWCA have the expected sign. However, the t value is low. The coefficient of FILT is also not significantly different from zero. Its negative sign may be due to the ambiguity of interpreting water filter expenditures mentioned above.

The coefficient estimates of SENS1 and SENS2 are both negative, which means that people who are likely to make relatively robust value judgements tend to have a relatively *high* WTP. The influence of these two variables on WTP is thus unambiguous, but again, the t values are only around unity.

A study of unusual observations was also undertaken. Three observations had a leverage exceeding $2 \times 12/106$. Exclusion of these three observations from the analysis caused only very small changes in coefficients and t values. The same is true for the case when PMYEAR instead of PMWCA was included as an explanatory variable.

To summarise, the estimation results are encouraging in the sense that the coefficient estimates of the two important economic variables PINCOM and PMWCA have the expected sign. In addition, SENS1 and SENS2 have an unambiguous influence on WTP. However, only a few of the coefficients are significantly different from zero. This lack of statistical significance implies that the respondents' answers to the binary choice questions are to a large extent unpredictable on the basis of the selected explanatory variables. Only the bid seems to be of real importance for the answer.

The relatively small sample size may be one reason for these weak results. However, another explanation may be that the valuation problem was too

Table 6.15. *ML estimation results for the LOGIT model*
$\text{BIDYES}_i = 1/[1 + \exp(-(\beta_0 - \beta_1\text{BIDLIRE}_i + \beta_2 -\text{SEX}_i + \beta_3\text{AGEYEAR}_i + \beta_4\text{EDUC}_i + \beta_5\text{CHILD}_i + \beta_6\text{SENS1}_i + \beta_7\text{SENS2}_i + \beta_8\text{PINCOM}_i + \beta_9\text{TWOPI}_i + \beta_{10}\text{FILT}_i + \beta_{11}\text{PMWCA}_i))]$, $i = 1, \ldots, 106$

Variable	Coefficient estimate	SE	t value [p]
Constant	−0.052906	2.198	−0.024 [0.98]
BIDLIRE	0.0015805	0.0003937	4.015 [0.00]
SEX	−0.14644	0.5702	−0.257 [0.80]
AGEYEAR	−0.0058614	0.02034	−0.288 [0.77]
EDUC	0.16201	0.3862	0.420 [0.67]
CHILD	1.3326	0.7637	1.745 [0.08]
SENS1	−0.73690	0.7702	−0.957 [0.34]
SENS2	−0.59947	0.5930	−1.011 [0.31]
PINCOM	0.000027500	0.0002977	0.924 [0.36]
TWOPI	−0.057982	0.2780	−0.209 [0.83]
FILT	−0.91096	0.7213	−1.263 [0.21]
PMWCA	0.0028230	0.003502	0.806 [0.42]

Log-likelihood $= -47.35$, pseudo-$R^2 = 0.34$, $\chi^2(11) = 48.46$, significance level of $\chi^2 = 0.000$

Actual and predicted outcome (frequency):

	Predicted		
Actual	0	1	Total
0 (no/don't know)	49	14	63
1 (yes)	11	32	43
Total	60	46	106

Note:
SE, standard error.

unfamiliar to the respondents. In this case, factors other than those usually included as explanatory variables in demand functions for market goods may be more important. For example, we have seen on pp. 155–7 that the bid proposed in the interview was an important determinant for the respondents' answers to the open-ended WTP questions.

In order to approach this familiarity/predictability issue, a most interesting extension of this work would be to use the interview data for an estimation of a demand function for mineral water, i.e. a market good which is familiar to the respondents, and to compare the predictability of this demand behaviour with the predictability of the WTP for the Water Resource Management Plan. A more thorough analysis would also include experiments with probability models other than the logit model.

Discussion

Evaluation of benefit estimates

It was noted on p. 131 that there is only a limited experience of the CVM in Italy. This study should thus best be seen as a pioneering piece of work rather than a straightforward application of the CVM. In addition, the valuation issue selected for the study is intricate in several respects. These circumstances explain why in the preceding sections we have made a rather detailed presentation of the institutional setting, the development of the survey instrument and the survey. These details help in evaluating the benefit estimates computed on pp. 160–1.

There are at least two features of the benefit estimates that deserve closer examination: (1) they appear to be high, and (2) they do not seem easy to predict by conventional socioeconomic variables. These features are discussed below.

Size of the benefit estimates

The judgement that the estimates are high follows readily from a comparison with income (mean WTP constitutes 1.2% of mean household income) and mineral water expenditures (mean WTP constitutes 166% of mean household expenditures). Moreover, other published CVM studies of drinking-water quality and groundwater resource protection, carried out elsewhere, revealed smaller mean WTPs, in the range of ITL 60 000– 195 000; see Appendix 3 of this chapter for a summary. However, a straightforward conclusion that the WTP amounts stated by the respondents are exaggerated does not follow from these comparisons. The comparisons suggest rather that the benefit estimates may not exactly concern the valuation issue that the respondents were asked to consider.

To summarise the valuation issue, the proposed Special Fund for a Groundwater Resource Management Plan encompasses two entities:

(1) *Groundwater protection*: an environmental resource and public good in terms of it acting as a barometer of wider environmental quality and being a source of drinking-water supplies.

(2) *Drinking water security*: a private good which is substitutable (to varying degrees) by water treatment techniques, at the level of the municipal supplier or via domestic filters, and bottled mineral water.

While 96% of respondents replied in the affirmative to the question on the importance of groundwater quality as a natural resource in its own right, a high valuation of the non-use benefits of groundwater resource protection *implied by the actual valuation scenario* does not seem a full explanation for the large WTP. However, the respondents might have perceived benefits other than those that were explicitly implied by the valuation scenario. In other words, there may be 'embedding effects'. Indeed, psychological studies have found that outcomes of decision-making processes are sensitive to the *context* in which problem is proposed; thus water-quality considerations may have included an element of 'symbolic demand', i.e. where WTP is used to demonstrate strength of feeling rather than being a true reflection of perceived benefit.

It is also possible that people's perceptions of the present health risks are quite different from the actual ones. Exaggerated media reports and mineral-water publicity compaigns have probably influenced people's perception of tap-water quality. The survey also showed that by far the most important motive for mineral-water consumption was tap-water substitution. Such underlying perceptions may have implied that simple *risk aversion* among subjects was a more influential factor than would be justified from a scientific viewpoint; the actual health risks are effectively tiny. It should also be noted, however, that specific questions regarding concern for water quality as a major environmental issue and specific risks connected with municipal water supply revealed relatively low levels of concern among respondents.[11]

Furthermore, 'water quality' constitutes a particularly complicated set of characteristics, only a subset of which are addressed by the proposed Resource Management Plan; water hardness and taste, for example, would remain the same in practice, although expected improvements in such aesthetic properties may have been included in individuals' valuation functions.

Finally, the WTP elicitation method's potential influence on stated WTP should not be neglected. Kriström (1993) noted that there is empirical evidence that binary choice WTP questions tend to result in higher mean WTP estimates than did open-ended ones. The reasons for this phenomenon still remain to be thoroughly studied. Compliance bias in the case of binary choice WTP questions may be present, but on the other hand, the binary choice setting may be more effective in revealing a respondent's *maximum* WTP.

To conclude, we find reasons to believe that the benefit estimates are not completely consistent with the valuation issue actually studied. In particular, the relatively large observed WTP is likely to reflect a situation when non-marginal health risks can be remedied by groundwater protection. This is not the situation in Milan, at least not at present.

Prediction of benefit estimates

The analysis on pp. 164–6 showed that our estimation of a WTP function was not very successful. While correct signs of the coefficient estimates of some important explanatory variables were an encouraging finding, the statistical significance was weak. This may partly be due to a relatively small sample size, but it is also likely that this indicates the presence of some problems related to the issues discussed above.

Perhaps a fundamental problem is that the change in the level of provision of the environmental good was basically marginal in this case, rendering the policy judgement more difficult (see Schelling, 1984). Related to this issue is a possible 'bench marking' factor. Given the rather broad resource description and scope of the proposed policy measure, it is possible that respondents had very different perceptions of the size of the change for which they would be paying.

If this were the case, it follows that predictions of WTP would require more information on these perceptions than was obtained by the survey used in this study. From the researchers' point of view, this study's application of the CVM to an intricate valuation issue has highlighted the importance of collecting a broad array of background information during a CVM survey for the purposes of consistency tests and interpreting answers in a more complete sociological context. When determining non-market preferences and evaluating externalities, such information constitutes a very important complement to the results gained from statistical treatment of numerical data obtained by WTP questions via the CVM.

A comparison with costs

As was explained in the beginning of this chapter, the background to this study was partly that an analysis of socially efficient contamination levels could not be undertaken due to lack of information on benefits. This case study has now provided some benefit estimates. However, the conclusions drawn above imply that they have to be used with great care, since it is not entirely clear for what benefits they are valid.

It can nevertheless be of interest when seen in the light of some reflections on the order of magnitude of the benefit estimates. For a case study of an area called Alto Vicentino, north of Vicenza in the Veneto region in northern Italy, Söderqvist (1994) estimated, *inter alia*, the costs of water purification by granulated active carbon (GAC). Such treatment of drinking water removes a range of different contaminants. The annual costs per capita for a GAC treatment of all water abstracted from the aquifer of Alto Vicentino were estimated to be about ITL3770 million/180 000 persons = ITL20 900. This means that the benefit estimates of this Milan study (ITL645 000 per household) would be valid also for the area of Alto Vicentino and for this treatment option, there would be reason to believe that the benefits of treated drinking water are higher than the costs.

However, this transfer of benefits to the situation in Alto Vicentino is completely speculative. The appropriate way to use the benefit estimates of this study is to make a comparison with the costs of realising a Management Plan of the kind outlined in the valuation scenario. To produce such cost estimates is a subject for future research.

Recommendations

On pp. 155–7, it was found that the bids suggested in the binary choice question served as anchors for the respondents' answers to the open-ended WTP questions. This means that, at least in this study, reliable additional information on WTP was not obtained by combining a binary choice WTP question with a subsequent open-ended question. We believe that the anchoring effect is due to the fact that the study concerned an intricate valuation issue for which a considerable degree of preference uncertainty existed among the respondents. The findings make this combination of WTP questions inappropriate.

This study was in many respects a pioneering application of the CVM in Italy. Many insights have been gained that may be useful for future Italian CVM studies. For example, it was concluded at an early stage that face-to-face interviews were the only reasonable data collection option. At present, the response rates that can be expected for postal questionnaires are too low, and telephone interviews are only likely to be useful for valuation issues that are already familiar to the respondents before the interview. Moreover, the face-to-face interviews turned out to work rather well in practice. A high response rate was obtained and no interviewer bias was found. The lack of predictability and other problems related to the benefit estimates cannot be blamed on the face-to-face interviews per se. These problems concern instead the types of data that were collected by the interviews.

Appendix 1: Water supply in Milan

Water quality

Milan's basically good-quality water can be attributed to: (1) large aquifer reserves; (2) flexibility in supply due to a high number of wells, and potential to 'dilute' more polluted water from certain wells with water from better ones; and (3) activated carbon treatment facilities in some distribution stations.

Milan's municipal waters are sourced entirely from natural aquifers underlying the city. Well technology in the area dates back to pre-Christian times, as Milan was home to ancient Roman populations. Specific literature references, however, begin in the thirteenth century when the convenience and freshness of readily available well water was remarked upon by Bonvesin de la Riva, a contemporary of Dante:

> In the city there are neither tanks nor pipes coming from far away but living waters, natural, wonderfully suited to being drunk by man, pure and wholesome, always at arm's reach, never scarce even in dry weather and so abundant that in every near-decent house there is nearly always a source of living water known as a well.

At that time, there were an estimated 6000 active wells. Over the centuries, improved scientific understanding revealed the hidden dangers of surface runoff contamination of shallower wells and infiltration from wastewater. Consequently, the option of piping in clean water all the way from sources near Bergamo, and abandoning the local aquifer supply, was seriously considered in the nineteenth century. Opposition from the city of Bergamo led Milan to reconsider the inherent value of its natural groundwater endowment and the ensuing programme of hygienic well construction, taking waters from a depth of over 100 m (Airoldi and Casati, 1989).

This practice has continued to the present day. Milan is divided into 31 distribution zones which together manage 582 wells (Provincia di Milano, 1993). The combination and number of wells in service at any one time is determined by water quality characteristics – the most seriously contaminated wells are closed completely while waters are mixed and diluted among operating wells to satisfy the legal standard for drinking-water quality.

Until spring 1994, when Milan's special exemption from the chlorinated hydrocarbon standards of DPR 236/88 expired, this type of resource management, broadly based on 'pollution prevention' (i.e. well closure) rather than treatment, sufficed. Activated carbon treatment units have since been installed in many distribution centres to reduce the levels of solvent-derived residues in drinking water. Despite an ongoing programme of well-water remediation, whereby waters in out-of-service wells are treated and then returned, so that the well can be put back in service, Milan's water supply remains under threat from serious contamination. Experts frequently debate alternative strategic options, including the piping-in of surface waters over long distances. As many of the lakes and rivers in northern Italy are also seriously compromised in terms of effluent controls, there is widespread consensus that the optimal solution, from both the economic and qualitative perspectives, remains local groundwater. Several protection measures still need to be implemented, however, including pollution control in the hydrological basin (acquifer infiltration zone) around Milan's aquifers, further water pre-treatment technology in the city itself and extraction from deeper layers. The experts' consensus formed the basis of the valuation issue adopted in the CVM survey.

Water consumption

Data on consumption are broken down in the Tables 6A.1 and 6A.2. While agriculture in the Milan province accounts for only a small proportion of annual groundwater withdrawals (ca 1%), this sector is a main contributor to pollutant load, principally nitrates and pesticides (Provincia di Milano, 1993). In other words, agricultural activities infringe on water availability in qualitative, rather than quantitative, ways.

Table 6A.3 shows a rough breakdown of household water uses, comparing Italy with Sweden and the UK. Serious leakages and relatively low tariff levels are significant contributors to the higher consumption level in Italy, as highlighted by the 'major losses' category. Drinking water is clearly a relatively moderate use category as far as municipal water supplies are concerned. Looked at qualitatively, from the public health perspective, tap-water quality is a prime concern.

Institutional framework and tariff systems

Water resources are managed by a high diversity of public structures in Italy, in highly fragmented systems in terms of technical, distribution and administrative functions. In the city of Milan, the *Acquedotto di Milano* is responsible for water supply and management in its technical capacity while another unit carries out the

Table 6A.1. *Classification of water supplies*

% annual consumption (1991)	Milan – city	Milan – province
Public (municipal network)	87	67
Industrial (registered private wells)	13	32
Agricultural	0	1

Table 6A.2. *Water consumption*

	Milan – city[a]	Milan – province[a]
Total water consumption (1991)		
m³/year	350 m	1014 m
population	1.4 m	3.9 m
m³/capita per year	255	260
Public water consumption (1991)		
m³/year	303.5 m	694 m
population	1.4 m	3.9 m
m³/capita per year	221	178
% public water use/capita	87	67

Note:
[a] m, million.
Source: Provincia di Milano (1993).

Table 6A.3. *Domestic consumption in some European countries*

Use (% breakdown)	Italy	Sweden	UK
Wastewaters, WC	27.3	20.0	35.4
Personal hygiene	22.7	38.5	34.6
Drinking, food, washing-up	13.6	22.0	13.8
Cleaning and clothes washing	13.6	11.8	10.0
Various other uses	13.6	6.2	6.2
Major losses	9.0	–	–
Total %	99.8	98.5	100
l/day	220	195	130

Source: Galli (1990).

administrative function. The lack of integration and synergy between two municipal bodies, combined with a generally low level of service provision in terms of water treatment, investment in pipelines and leak control, account for major system inefficiencies. This is exacerbated by low tariff levels. At a national level, average water rates in Italy are around ITL600/m³, compared with ITL2400/m³ in Germany and ITL1700/m³ in France; this heavy subsidisation means that it is often cheaper to waste water if a leak develops than to call a plumber (Galli, 1990).

Tariffs are generally structured to increase with water consumption rate in the domestic sector, ranging from ITL170–420/m³. Municipal water for agricultural uses is heavily subsidised at ITL100/m³ and industrial uses are priced at the uniform maximum rate of ITL420/m³. Considering use rates and average sectoral distribution, the average water tariff in Milan is ITL250/m³ (or ITL60 000–100 000/year per household). Within the civil sector, it is not possible to distinguish between domestic and non-domestic water consumption as tariffs are set on a building-by-building basis and many buildings house more than one type of activity, shops and craftsmen occupying the lower floors and apartments above.

In line with the building-level classification, there is also normally a single water meter. Individual tariffs are divided according to water consumption rate, calculated as an average according to the number of units (apartments) in the building. Normally, the building administrator divides the water bill among occupants on the basis of unit size. While water bills are issued every two months, payment at the household level (to the building's administration) is normally annual and incorporated within a universal building services account, along with cleaning and lighting of common areas, structural repairs etc. In this context, the average annual household water costs of ITL60 000–100 000 is not particularly distinctive, hence the observed lack of awareness of the 'cost' of water.

Appendix 2: Final Questionnaire (translation)

European Science Foundation/Fondazione Eni Enrico Mattei Contingent Valuation Survey for the Economic Valuation of Water Quality

Description of the research project

The aim of this study is to better understand the attitudes and preferences of Milan residents with respect to the quality of drinking water, and water resources more generally, as well as to estimate the full economic value of water as a resource. The study is purely academic; there is no involvement whatsoever of the city administration or indeed any other public or private body. The *European Science Foundation* and the *Fondazione Eni Enrico Mattei* (a Milan-based institute specialising in environmental, development and energy economics founded in 1989) have entrusted researchers from the Università di Milano with the execution of the survey.

1. **Respondent's address and water classification zone**

   Socio-economic information

2. **Sex**   m/f
3. **Age** $< 20/20–25/26–30/31–40/41–50/51–60/ > 60$
4. **Education**

   none/primary school/high school (intermediate level)/high school (diploma)/university degree

5. **Professional status**

   entrepreneur/freelance/shopkeeper, craftsman/manager, university professor/white collar worker, teacher/specialised workman, technician/workman/support worker/unemployed, in search of first job/housewife/student/other (to be specified)

6. **Number of people in household**
7. **Family structure**

   single/couple/couple with child/couple with children and other relatives/single parent with child or children/group of sharers (e.g. students)/other

8. **Do you (respondent) contribute regularly to household income?**
9. **Do you routinely do the grocery shopping?**

   Water consumption

10. **Do you use tapwater for ...?**

    drinking water/cooking (including tea and coffee)

11. **What is your general opinion of Milan's tapwater?**

    excellent/good/satisfactory/bad/terrible

12. **In particular, do you think that the water in Milan ...?**

    is bad tasting/is polluted/carries health risks

13. **Do you personally drink mineral water (on a regular basis)?**

    yes/no

    If yes, what are your two main motives?

    taste and/or fizziness/tapwater substitute/specific health or therapeutic properties/substitute for other soft or alcoholic drinks/habit

14. **Have you installed a domestic water filter**

    yes/no

    If yes ...?

    since when?

    how much does it cost? Initial investment? Annual maintenance costs?

    what are its particular advantages/functions?

Scenario description

I Milan's drinking water is sourced entirely from underground acquifers.

II Current drinking-water supply in the city (only just) complies with standards set by the European Commission for water quality.

III Production activities, both industrial and agricultural, tend to have a negative impact on groundwater quality which necessitates the introduction of a new, integrated and extensive Water Resource Management Plan relating to Milan.

IV *Doing nothing*, on the other hand, would lead to a continued rise in the concentration of dangerous substances implicating health risks, worsening of general environmental quality...

V The implementation of the Management Plan, however, involves additional costs to which Milan residents would have to contribute.

15. **In the event of a referendum concerning the introduction of a Groundwater Resource Management Programme, would you be in favour of the establishment of a *Special Fund* for resource management that implicates household contributions?**

    YES → Would you be willing to pay a household contribution of ITL*X*?
    YES → Would you be willing to pay any more?
          YES → How much altogether? ITL ____
          NO
          DON'T KNOW
    NO → How much would you be willing to pay between 0 and the sum mentioned?
          ITL ____. *If nothing*, why?
    DON'T KNOW

    NO → Would you instead prefer...?
    – increased *water tariffs* → YES/NO
    – increased regulation and operating restrictions on producing activities which would bring about general increases in *the price of consumer goods* → YES/NO
    If NO to both, **why are you unwilling to dedicate more resources to water resource protection?**
    DON'T KNOW

Water consumption behaviour and opinions

16. **Does your family drink mineral water?**
    yes/no
    If yes...?
    How much do you spend on average for a bottle of mineral water?
    ITL ____ /don't know
    Size of bottle?
    0.5/1/1.5/2 litres

Average household weekly consumption?

___ litres

Mineral water preference

still/fizzy?

17. **Do you think that tapwater and mineral water may be considered direct substitutes?**

yes/no/don't know

18. **Do you believe that groundwater is an important resource to protect, nevertheless, as regards environmental protection in general?**

yes/no/don't know

19. **In your opinion, which are the two most pressing environmental concerns?**

atmospheric pollution

poor-quality drinking water

urban traffic and noise levels

waste disposal

global issues (climate change, stratospheric ozone depletion)

environmental degradation in general (rivers, oceans, landscapes etc.)

20. **In the light of these considerations, would you like to modify your expressed WTP?**

yes/no

If yes, what is your new WTP?

ITL___/don't know

General information

21. **Type of home**

single- or two-family house/apartment

rented/owned/other

22. **Gross household annual income**

< ITL10   million/11–20 m/21–30 m.31–50 m/51–70 m/71–100 m/ > 100 m/no answer

23. **How much do you currently pay in water tariffs?**

ITL___/don't know

Final questions for interviewer

Interviewer's identification code

Duration of interview   ___ minutes

Was the interviewee particularly   interested/helpful/well informed?

During the interview, did the interviewee comment on who holds main responsibility for water quality?

citizens

public institutions

industry

agriculture

other (to be specified)

Appendix 3: Summary of selected similar CVM studies

Reference	Object, study	Survey method	Estimated WTP	Response rate	Comments
Kwak and Russell (1994)	Increased water security to be implemented via a Government Plan to reduce the probability of major contamination incidents to zero or near zero. Study locus: Seoul, South Korea	Face-to-face interviews. Payment vehicle: combination of increased water tariffs and income tax. Elicitation: payment card	Mean: US$40 per household and year (0.25% of average income)	298 observations	WTP function with strong explanatory variable was estimated
Edwards (1988)	Prevention of uncertain, future nitrate contamination of groundwaters used for drinking water supply. Study locus: Cape Cod (Massachusetts), USA	Mail questionnaire. Payment vehicle: bond, via public referendum. Elicitation: bid variable	Mean varies according to scenario presented; 10 different versions	1000 households in sample; 78.5% replies, of which 58.5% provided sufficient information	Highlights effects of demand and supply uncertainty on WTP; strong influence of bequest motives on total WTP; small option value cf. option price. Powerful WTP function: strong income effect; price of bottled water not significant
Power et al.	Increased water supply protection. Study locus: 12 communities in north-eastern USA	Mail questionnaire. Payment vehicle: water tariffs. Elicitation: payment card (US$ 0–350)	Mean: US$ 61.55 per household and year	Response rate 50%, yielding 1006 responses	Categorical regression analysis revealed: (a) significant WTP predictors: income, contamination incident experience, water safety perception, number of perceived contamination sources, public/private water supply, bottled water expenditure

					(b) non-significant predictors: sex, age, children, knowledge of groundwater issues
Schultz and Lindsay (1990)	Hypothetical groundwater protection plan Study locus: Dover (New Hampshire), USA	Mail questionnaire Payment vehicle: property taxes Elicitation: bid variable (US$1–500)	Mean US$ 129 per household and year Median: US$ 40	Questionnaire sent to 600 homes; 59.3% response rate	(a) Significant WTP predictors: income, assessed land value (+); age (−) (b) Non-significant predictors: residence time in the community, sex, education, children, knowledge of groundwater issues
Jordan and Elnagheeb (1993)	Improved drinking-water quality (and perception of potential groundwater contamination) Study locus: Georgia, USA	Mail questionnaire Payment vehicle: water tariffs Elicitation: payment card	Median: US$ 790.60/annum for public supplies; US$ 1062.70 for private well users	567 in sample, 35% response rate (192) completed questionnaires	(a) Significant WTP predictors: income, education, water quality uncertainty, women, blacks (+); age (−) (b) Non-significant predictors: residence time in the community, sex, education, children, knowledge of groundwater issues

Notes

1 We have worked together with Marcella Pavan and Paolo Frassetti at different stages of this study. We are grateful for their efforts, and we also thank Lars Bergman, Carolina Sbriscia Fioretti, Per-Olov Johansson, Bengt Kriström, Richard Lloyd, Robin Mason, D. Michael Pugh, Timothy M. Swanson, and Marco Vighi for help and comments. The usual disclaimer applies. Financial support from the European Science Foundation and Fondazione Eni Enrico Mattei is gratefully acknowledged.

2 The distinction between pollutant and contaminant is merely quantitative. A contaminating substance becomes polluting once it reaches a concentration at which the risks of causing damage are significant. Of course, 'risk' and 'significant' are value-laden terms, the meaning of which depends on subjective assessments as well as the state of scientific knowledge.

3 See, for example, Boadway and Bruce (1984) for a thorough exposition, and Johansson (1991) for an introduction.

4 See Johansson *et al.* (1989) on the mean versus median issue.

5 All monetary amounts in this paper are quoted in 1994 prices, if not otherwise stated. ITL1000 ≈ US$0.7.

6 Some interpretation subjectivity remains, however, such that a respondent who never makes use of the resource in question (or something analogous to it) may have a justifiably high WTP if, to him or her, the non-use values are particularly important. Conversely, the high WTP of a person who is not a user of the resource may be taken as symptomatic of an unreliable methodology.

7 Lack of demand for additional explanatory information (mentioned on pp. 134–7) contrasts with this observation, highlighting the difficulties involved in seeking clear marketplace-type decisions for environmental goods.

8 The standard errors are approximations computed by the WALD command of Limdep 6.0 (see Greene, 1991, pp. 156–9).

9 Observation with outside values are defined as those located to the right (left) of the third quartile plus 1.5 times the distance between the third and first quartiles (the first quartile minus 1.5 distances). In the case of far outside values, the distance is multiplied by 3 instead (see, for example, Körner *et al.*, 1984). In this case, there are no outside values or far outside values in the lower tail of the distribution of BIDINC, and there are no far outside values in the upper tail of the distribution.

10 That is, MODIFWTP is ITL100 000. Note that for this respondent, anchoring effects of the kind described above are not present, since he did not accept the Fund and thus was not proposed any bid.

11 Interviewers were asked to complete some specific questions about interviewee behaviour, at the end of each questionnaire. While there was a good level of cooperation amongst respondents (98%), fewer than 20% displayed positive interest and only 2% were deemed to be well informed about the subject.

References

Airoldi, R. and Casati, P. (1989). *Le Falde Idriche del Sottosuolo di Milano*. Commune di Milano.

Ayer, M., Brunk, H. D., Ewing, G. M. and Silverman, E. (1955). An empirical distribution function for sampling with incomplete information. *Annals of Mathematical Statistics*, **26**, 641–7.

Bergman, L. and Pugh, D. M. (1994). *Environmental Toxicology, Economics and Institution: The Atrazine Case Study*. Kluwer Academic Publishers, Dordrecht.

Bishop, R. C. and Heberlein, T. A. (1979). Measuring values of extra market goods: are indirect measures biased? *American Journal of Agricultural Economics*, **61**, 926–30.

Boadway, R. W. and Bruce, N. (1984). *Welfare Economics*. Basil Blackwell, Oxford.

Braden, J. B. and Kolstad, C. D. (eds.) (1991). *Measuring the Demand for Environmental Quality*. North-Holland, Amsterdam.

Databank (1993). *Acque Minerali* (market research report). Databank srl, Milan.

De Bernardi, M., Zanasi, A. and Brazzorotto, C. (1993). *Guida Ufficiale delle Acque Minerali Italiane in Bottiglia*. Masetti Editore, Bologna.

Diamond, P. A. and Hausman, J. A. (1994). Contingent valuation: is some number better than no number? *Journal of Economic Perspectives*, **8**, 45–64.

Edwards, S. F. (1988). Option prices for groundwater protection. *Journal of Environmental Economics and Management*, **15**, 475–87.

Freeman, A. M. III (1993). *The Measurement of Environmental and Resource Values: Theory and Methods*. Resources for the Future, Washington, DC.

Galli, G. (1990). *Imprese Ambiente* 1990/5.

Greene, W. H. (199). *Limdep. User's Manual and Reference Guide*. Econometric Software, Inc. Bellport, NY.

Hanemann, W. M. (1994). Valuing the environment through contingent valuation. *Journal of Economic Perspectives*, **8**, 19–43.

Johansson, P.-O. (1991). *An Introduction to Modern Welfare Economics*. Cambridge University Press, Cambridge.

Johansson, P.-O. (1993). *Cost–Benefit Analysis of Environmental Change*. Cambridge University Press, Cambridge.

Johansson, P.-O., Kriström, B. and Mäler, K.-G. (1989). Welfare evaluations in contingent valuation experiments with discreet response data: comment. *American Journal of Agricultural Economics*, **71**, 1054–6.

Jordan, J. L. and Elnagheeb, A. H.(1993). Willingness to pay for improvements in drinking water quality. *Water Resource Research*, **29**, 237–45.

Kriström, B. (1990). *Valuing Environmental Benefits Using the Contingent Valuation Method*. Umeå Economic Studies, no. 219, University of Umeå.

Kriström, B. (1993). Comparing continuous and discreet contingent valuation questions. *Environmental and Resource Economics*, **3**, 63–71.

Körner, S., Ek, L. and Berg, S. (1984). *Deskriptiv Statistik*. Studentlitteratur, Lund.

182   J. Press and T. Söderqvist

Kwak, S. J. and Russell, C. S. (1994). Contingent valuation in Korean environmental
planning: a pilot application to the protection of drinking water quality in Seoul.
Environmental and Resource Economics, **4**, 511–26.

Larson, D. M. (1992). Further results on willingness to pay for nonmarket goods.
Journal of Environmental Economics and Management, **23**, 101–22.

McClelland, G. H., Schulze, W. D., Lazo, J. K., Waldman, D. M., Doyle, J. K., Elliot, S. R.
and Irwin, J. R. (1992). Methods for measuring non-use values: a contingent
valuation study of groundwater cleanup. (Mimeo) Center for Economic
Analysis, University of Colorado. USEPA no. CR-815183.

Mitchell, R. C. and Carson, R. T. (1989). *Using Surveys for Valuing Public Goods: The
Contingent Valuation Method*. Resources for the Future, Washington, DC.

Ng, Y.-K. (1979). *Welfare Economics: Introduction and Development of Basic
Concepts*. Macmillan, London.

Portney, P. (1994). The contingent valuation debate: why economists should care.
Journal of Economic Perspectives, **8**, 3–17.

Power, J. R., McClintock, C. and Alle, D. J. (1991). The impact of contingent information on water supply protection. In *Proceedings of a Conference of the European
Association of Environmental and Resource Economists*.

Provincia di Milano S.I.F. (1993). *Prelievi Idrici da Falda nel Milanese*, Milan.

Schelling, T. C. (1984). *Economics of Reasoning and the Ethics of Policy in Choice and
Consequence*. Harvard University Press, Cambridge, MA.

Schultz, S. D. and Lindsay, B. E. (1990). The willingness to pay for groundwater protection. *Water Resources Research*, **26**, 1869–1875.

Smith, V. K. and Desvousges, W. H. (1986). *Measuring Water Quality Benefits*. Kluwer-
Nijhoff Publishing, Boston, MA.

Söderqvist, T. (1994). The costs of meeting a drinking water quality standard: the case
of atrazine in Italy – *The Atrazine Case Study*, ed. L. Bergman and D. M. Pugh, pp.
151–71. *Environmental Toxicology, Economics and Institutions*, Kluwer
Academic Publishers, Dordrecht.

Söderqvist, T. (1995). Estimating the benefits of health risk reduction: the contingent
valuation method applied to the case of residential radon radiation. *Beijer
Discussion Paper Series* no. 57. Beijer International Institute of Ecological
Economics, Stockholm.

Söderqvist, T., Bergman, L. and Johansson, P.-O. (1995). An economic approach. In
Pesticide Risk in Groundwater, ed. M. Vighi and E. Funari, pp. 241–58. Lewis
Publishers, Boca Raton, Fl.

Vighi, M. and Zanin, G. (1994). Agronomic and ecotoxicological aspects of herbicide
contamination of groundwater in Italy. In Environmental Toxicology,
Economics and Institutions – *The Atrazine Cast Study*, ed. L. Bergman and D. M.
Pugh, pp. 111–39. Kluwer Academic Publishers, Dordrecht.

Part III

The analysis of market and regulatory failure

7   Market failure[1]

Robin Mason

Introduction

This chapter examines the reasons why private decision-making on the part of agricultural chemical manufacturers might lead to products with socially undesirable characteristics being sold in socially inefficient quantities.[2] The source of the problem is *market failure*: some significant economic factor goes 'unpriced', so that economic activity is undertaken without consideration of its full impact. Two market failures will be considered. The first is the presence of externalities (both the standard story of a damage externality, in which agents considered only the private, and not social, costs of their actions; and a 'surplus externality', arising because profit-maximising firms consider only the marginal consumer, instead of all consumers, in their production decisions). The second is imperfect competition – the effect of the structure of the agriculture chemical manufacturing industry (a small number of large multinationals competing in the same product market) on producers' choices of chemical characteristics.

Externalities

An externality arises when the decisions of some economic agents (individuals, firms, governments) – whether in production, in consumption, or in exchange – affect other economic agents, and are not included in the priced system of commodities, i.e. they are not compensated. An alternative way of expressing this problem is to say that *property rights* are not assigned appropriately, so that the incidence of effect does not coincide with the distribution of legally recognised controls. A third equivalent statement is that economic agents consider only their private marginal costs when making decisions; they do not consider the total, or social (marginal), costs of their

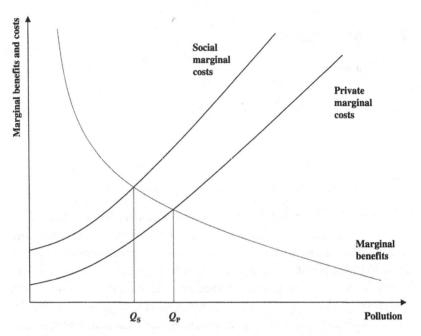

Fig. 7.1. Social and private marginal benefit and cost schedules.

actions. It is the gap between private and social marginal costs that gives rise to the externality.

Figure 7.1 illustrates the general case of a private agent undertaking some activity which generates a private benefit (for example, profit from running a firm), but which also gives rise to a cost due to the pollution that it creates. The horizontal axis measures the amount of pollution; the vertical axis measures the marginal benefits and costs from the activity. Assume, for simplicity, that private and social benefits from the activity are the same. Total benefits are assumed to increase with the amount of pollution (because, for example, the firm's output and the amount of pollution generated are directly correlated), but at a decreasing rate; the marginal benefit is therefore positive, but decreasing with the quantity of pollution. Now assume that private and social costs differ; in particular, suppose that private costs are *lower* than social costs – that is, the pollution inflicts less damage on the agent who creates it than it does on society as a whole. An example might clarify this point: imagine a farmer applying pesticide to her[3] crops. The farmer faces some health risk in doing this – she might inhale some of the chemical, it might come in contact with her skin, etc. She does not face, however, the full cost of her actions; she may not, for example, care about the health of people whose drinking water comes from groundwater reservoirs lying beneath her land. Total costs (both private and social) increase with the amount of pollution at an increasing rate; marginal cost is therefore positive and increasing in the quantity of pollution. The private optimal quantity of pollution is that level at which marginal benefit equals private marginal cost; this is denoted Q_p. Contrast this level with the socially optimal Q_s, at which marginal benefit equals social marginal cost. The private amount exceeds the social amount, since the agent does not face the full cost of her actions.

The nature of the market failure is the non-existence of a market for a certain commodity (in this case, the pollution generated by the activity of the agent). This causes a distortion in the market's allocation of resources, so that, in the presence of externalities, the market outcome does not maximise social welfare.

The effect of externalities on manufacturers' choices of product characteristics can be analysed in the following simple model[4]:

Consumers' demand

Consumers of an agricultural chemical care about several aspects of the product. A survey by Infomark (1993), on behalf of Du Pont, reported the following criteria used by farmers when choosing among herbicide treatments:

Experience with the product.
Low cost.
Absence of toxic residues in soil.
Selectivity.
Wide range of action.
Persistence.

It is clear from this list that consumers view products as combinations of characteristics; the demand for a certain product is demand for the particular combination offered by that product.[5] (This view of a good was first proposed by Lancaster (1966).) Three characteristics are of particular importance for this analysis: price, persistence, and the absence of toxic residues. Persistence indicates a chemical's ability to accumulate in the relevant environmental compartment – water for herbicides, animal fat for insecticides. It is particularly important therefore as a measure of the effectiveness of a chemical. Herbicides are required to be soluble in water, in order to be transported from the surface (where they are applied) to the roots of crops (where they act to kill weed seeds) by rain water. They are also required to have a long half-life in water, so that they persist for long enough in the soil to kill the weeds. A similar story applies to insecticides: they must be soluble in fat, so that they are absorbed into the tissues of insects (rather than expelled as waste); once there they must persist for long enough to ensure the death of the insect.

There are factors, however, which will counteract the desirability of high chemical persistence. It is often important that chemicals applied leave no toxic residues for when further crops are planted in rotation schemes. Moreover, farmers' concern for personal safety during the application of agricultural chemicals may reduce demand for very persistent chemicals. Both factors suggest that, while chemical persistence may be desirable at lower levels of persistence, the same characteristic becomes a 'bad' when it rises above some critical level.

The following notation will be used: the price p of the chemical is a function of the total quantity sold, Q, and an index A, representing the reliability of the chemical; this index will be referred to as the 'accumulation constant' of the chemical. The inverse demand curve can therefore be written as $p(Q, A)$. Consumers' preferences for chemical persistence mean that marginal utility (and hence price) is increasing in the level of this characteristic, up to some critical point $\tilde{A}$; after this point, concerns for toxic residues and personal safety mean that marginal utility/price decreases. Therefore: $\partial p / \partial A = p_A > 0$ for $A < \tilde{A}$; but $p_A < 0$ for $A > \tilde{A}$; and $p_{AA} < 0$. As usual the price of the chemical decreases with the aggregate quantity sold (i.e. $\partial p / \partial Q = p_Q < 0$).

Manufacturers' profit maximisation problem

Manufacturers must choose output Q and the accumulation constant of the product A to maximise profits, i.e. revenue $Qp(Q, A)$ minus costs $C(Q, A)$. The latter inceases with both quantity and accumulation, i.e. it is costly for manufacturers to produce either a greater quantity of the chemical (of a given accumulation constant), or a chemical of greater reliability (at a given quantity). A perfectly competitive industry will choose Q and A according to marginal pricing principles (this is a standard result in economics)[7]:

$$\max_{\{Q, A\}} Q p(Q, A) - C(Q, A)$$

$$p(Q^C, A^C) - C_Q(Q^C, A^C) = 0 \tag{7.1}$$

$$Q^C p_A(Q^C, A^C) - C_A(Q^C, A^C) = 0 \tag{7.2}$$

(where subscripts denote partial derivatives, e.g. C_Q is marginal cost with respect to quantity produced; p_A is the marginal valuation of a unit increase in the accumulation constant). Note that equation (7.2) implies that $A^C < \tilde{A}$, since $C_A > 0$; in other words, profit-maximising firms will choose an accumulation constant in the range where price is increasing in the chemical's persistence. These two equations will be considered further below, once the socially optimal case has been analysed.

Social planner's maximisation problem

Economists measure the benefit derived from an activity by the surplus that it generates. This surplus is defined in a very specific way – it is the marginal benefit received by consumers from purchasing the product, summed over all consumers, plus the profits made by firms selling the product. Equivalently, this surplus is equal to the area under the inverse demand curve $p(Q, A)$, i.e. the area marked CS in Figure 7.2. (Zero costs have been assumed in this figure for simplicity.) In the notation that has been used in this section, the surplus is given by $\int_0^Q p(q, A) dq - C(Q, A)$.

The object of the analysis now is to posit the existence of a 'social planner' – an omnipotent being who controls all activities in the economy for the good of society – and compare the choices of this planner with those of the competitive industry. The social planner must maximise the surplus W minus any social damages (such as health risks) that arise from the build-up of chemicals in the environment (for example, herbicides in the groundwater). Assuming that this damage can be quantified, denote it by the function D, which increases with both the quantity used and persistence of the chemical. The social planner maximises its objective by choosing output and the accumulation constant to satisfy two marginal equalities[8]:

$$\max_{\{Q, A\}} W(Q, A) = \int_0^Q p(q, A) dq - C(Q, A) - D(Q, A).$$

$$p(Q^S, A^S) - C_Q(Q^S, A^S) - D_Q(Q^S, A^S) = 0 \tag{7.3}$$

$$\int_0^{Q^S} p_A(q, A^S) dq - C_A(Q^S, A^S) - D_A(Q^S, A^S) = 0 \tag{7.4}$$

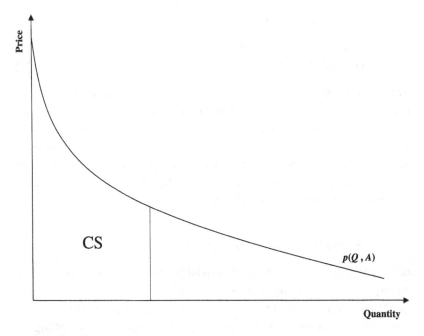

Fig. 7.2. Gross surplus.

Comparison of equations (7.1) to (7.4) shows that private profit maxi-misation by a perfectly competitive industry causes two inefficiencies. The first is the usual 'damage externality' – (other things being equal) firms sell too much of an excessively persistent chemical because they do not consider the social damages D caused by its use and accumulation in the environment. Hence, no term involving D appears in the first two equations, while marginal damages D_Q and D_A appear in the latter equations. In familiar language, firms making private production decisions fail to internalise the full cost of their actions, and so impose an externality on society.

The second inefficiency is more subtle, and involves a distortion in the persistence of the chemical produced by profit-maximisation firms. Equations (7.2) and (7.4) show that, other things being equal, if:

$$p_A(Q, A) \geq \frac{1}{Q} \int_0^Q p_A(q, A)\mathrm{d}q \qquad (7.5)$$

then the competitive industry will produce chemicals with socially exces-sive levels of persistence (the market accumulation constant A^c is greater

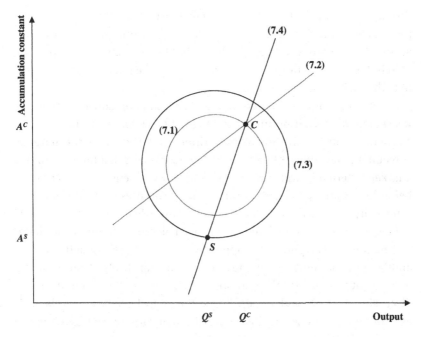

Fig. 7.3. Market and socially optimal solution.

than the socially optimal constant A^s). (Equation (7.5) is a sufficient condition for this to hold). The nature of the externality in this situation is that the competitive industry considers only the marginal consumer in its profit-maximising decisions; the social planner considers all consumers in choosing the accumulation characteristics of the chemical. This 'surplus externality' occurs on the benefit, rather than the cost, side of the equation.

Figure 7.3 illustrates these points graphically. The horizontal axis measures output of the industry, the vertical axis the accumulation constant of the industry's product. To simplify matters, it has been assumed that production of the chemical is a constant returns-to-scale (CRS) technology. This means that equation (7.1) is equivalent to a zero profit condition $Qp(Q, A) - C(Q, A) = \pi = 0$. The zero profit schedule in Figure 7.3 is the inner of the two circles. The schedule corresponding to equation (7.3) is concentric with this circle, but lies outside of it due to the additional term D_Q. The market and social solutions are then the intersection points of these schedules and the schedules corresponding to equations (7.2) and (7.4), respectively. The (7.2) schedule intersects the (7.1) line at the point C, giving the market output Q^C and accumulation constant A^C. The (7.4) schedule passes through this point, but intersects the schedule relating to equation (7.3) to the south-west of C, at point S, giving the socially optimal output Q^S and accumulation constant A^S. Notice that it has been assumed that equation (7.5) holds in this figure, so that both Q^C and A^C are unambiguously larger than the socially optimal levels.

Finally, is it possible that the market accumulation constant might be *lower* than the socially optimal level (might the inequality in equation (7.5) run in the opposite direction to that written)? If the benefit to the marginal consumer is *less* than the benefit to intramarginal consumers, then this 'market' externality will lead to the competitive accumulation constant being less than the socially optimal level. The presence of a 'damage' externality (where persistence of the chemical leads to environmental damages which are considered by the social planner but not by the chemical manufacturers) pulls in the opposite direction. The end result is *a priori* ambiguous, and the question must be settled empirically.[9] Unfortunately, data do not exist to test this condition. I will assume in the remainder of the discussion that Q^C and A^C are larger the socially optimal levels. This assumption is less severe the larger the damages caused by chemical accumulation (specifically, the larger is D_A).

Imperfect competition

Market failure due to imperfect competition arises when a small number of large firms compete in the same product market. The standard economic paradigm is *perfect competition*, in which each firm in an industry has such a small market share that it is unable to affect the price of its product[10] or the behaviour of its rivals. Social welfare is maximised in this case (for a proof of this statement, see Varian, 1992).

In contrast, when the number of firms is small, they realise their ability to improve their trading environment either by restricting their output (so that the price of their product rises), or by causing their rivals to decrease production. The welfare consequences of this behaviour are shown in Figure 7.4 for the simplest case of a single firm (a monopolist) influencing the price of its product by altering its output. The horizontal axis measures quantity of product sold; the vertical axis measures price, marginal revenue and marginal cost. If the industry were perfectly competitive, the firm would treat the price as fixed, and the profit-maximising choice of output would equate price and marginal cost. When the firm realises that it can

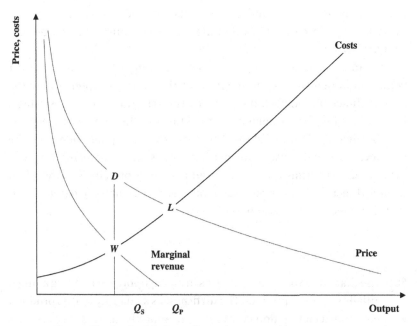

Fig. 7.4. The deadweight loss of imperfect competition.

alter the price it faces, its profit-maximising decision becomes to equate *marginal revenue* and marginal cost. Price, as usual, decreases with quantity (i.e. in order to sell a larger quantity, the firm must lower its price); marginal cost increases with quantity (i.e. total cost increases at an increasing rate with quantity; this is a standard assumption). It is easy to show that, for most forms of product demand, marginal revenue decreases with quantity sold, but lies below the price.[11] Figure 7.4 shows, therefore, that the monopolist chooses a lower output, gaining a higher price, than in the case of perfect competition. The result is a loss of social welfare equal to the area of the triangle marked *DWL* in the figure.

The agricultural chemical industry is an *oligopoly* – 80% of worldwide agrochemical sales are accounted for by 12 firms (see Nadai, 1995). This fact suggests that imperfect competition (where manufacturers do not treat price and/or the behaviour of competitors as exogenous) will be an important factor in determining the accumulation characteristics of these products. This section analyses the effect of market structure on manufacturers' choice of product accumulation characteristics.[12] The analysis will, unfortunately, be more technical than that of the previous section, a fact made unavoidable by the more difficult concepts involved. Readers who are less interested in the derivations, and are concerned only with the results, should turn to p. 198 for the conclusions.

The story is more complicated than the simplest case of imperfect competition that was told in the introduction to this chapter (where the output choice of a single monopolistic firm was examined). The strategic interactions between a small number of firms in the same industry must now be included in the analysis. This requires a set-up different from the one developed in the discussion of externalities: the demand for the products of different firms must be derived from first principles, rather than dealing directly with an inverse demand curve for a single good (the reason for this should become clear below).

Model basics

Suppose that there are n different firms in an oligopoly, each selling a single (potentially different) product. The ith firm sells a quantity q_i of its product, which has an accumulation constant a_i. To simplify matters, assume that there is a single, representative consumer who must decide which products, and how much of each, to buy. The consumer gains a gross surplus (before considering the cost) from the ith product of $\phi(q_i, a_i)$, and so gains a total gross surplus from all products purchased of $s = \sum_{i=1}^{n} \phi(q_i, a_i)$. This total gross surplus gives the consumer a utility of $u = G(s)$. The consumer prefers more to less in all aspects (i.e. for each product, the consumer prefers greater quantity to less, and a higher accumulation constant to a lower), but at a decreasing rate (for example, as the consumer receives more of one product, her desire for a greater quantity lessens)[13].

The consumer must choose the quantity of each product x_i to buy, given her preferences for the different goods, and given her income I. Product i costs p_i; so the consumer's problem is:

$$\max_{\{q_i\}} u = G\left(\sum_{i=1}^{n} \phi(q_i, a_i)\right)$$

$$\text{subject to } I \geq \sum_{i=1}^{n} p_i q_i$$

From this consumer's utility maximisation problem, the inverse demand function p_i for each product can be worked out:

$$p_i = \frac{\partial \phi(q_i, a_i)}{\partial q} \, G'(s). \tag{7.6}$$

(In future, partial derivatives will be denoted by subscripts, e.g. ϕ_q.) Equation (7.6) indicates why the more complicated structure of the problem has been adopted. The price of any product i is affected by the quantity and accumulation constant of that product (through the term $\phi(q_i, a_i)$). But it is also affected by the sales of other firms' products, through the term $G'(\Sigma_{j=1}^{n} \phi(q_j, a_j))$. This latter factor was not present in the model of the previous section and is the key difference which will allow the study of the effect of oligopolistic interaction.

Attention is now turned to firms' production decisions. Firm i will seek to maximise its profit, which is equal to revenues $q_i p_i$ minus costs $c(q_i, a_i)$, with costs increasing in both quantity and the product's accumulation constant. The analysis will be made considerably easier by assuming identical firms and products. In this case, $s = n\phi(q, a)$, $p = \phi_q G'(s)$ and $c = c(q, a)$. As before, the firms' profit-maximising decisions will be compared to those that would be made by a fictitious social planner, whose optimisation problem is considered first.

The social optimum

The social planner seeks to maximise the sum of total surplus from purchases of the products minus the costs of production, by choosing quantities purchased (q) and the accumulation constant of these products (a). Entry occurs freely until firms' profits are driven to zero; this is known as monopolistic competition and is analysed in the papers by Spence (1977) and Dixit (1979). The social planner's problem is then:

$$\max_{\{q,a,n\}} G(n\phi(q, a)) - nc(q, a)$$

In words: the social planner must choose output, the accumulation constant and the number of firms (and therefore the number of products on offer) to maximise social surplus. (Damages from the accumulation of the product in the environment will be included later, once the effect of imperfect competition alone has been determined.) Two first-order conditions describe the social optimum:

$$G'(s) = M(a) \tag{7.7}$$
$$M'(a) = 0 \tag{7.8}$$

where $M(a)$ is the minimum value of c/ϕ, the cost per unit of surplus of each product.

Firms' profit-maximising decisions

In this section, the social optimum is compared to the oligopolistic outcome. Firms compete by choosing the quantity to be sold and the accumulation constant of their product; the price they receive for this product is then given by the inverse demand function in equation (7.6). The problem of each firm is to:

$$\max_{\{q,a\}} qp - c = q\phi_q(q, a)G'(s) - c(q, a)$$

The number of firms in equilibrium is determined by the zero profit condition: $q\phi_q(q, a)G'(s) - c(q, a) = 0$. Suppose (as in Dixit, 1979) that the resulting n is sufficiently large that each firm does not consider the effect of its production decisions on total surplus s. Also, assume that the function ϕ is iso-elastic (this will simplify the analysis considerably):

$$\phi(q, a) = A(a)q^{\alpha(a)}$$

where $0 < \alpha < 1$. Then $1/[1 - \alpha(a)]$ is the price elasticity of demand; if $\alpha'(a) > 0$, then the price elasticity increases with a. The first-order conditions describing the profit-maximising outcome are:

$$G'(s) = \frac{M(a)}{\alpha(a)} \tag{7.9}$$

$$\frac{M'(a)}{M(a)} = \frac{\alpha'(a)}{\alpha(a)} \tag{7.10}$$

Comparison of equations (7.7) with (7.8) and (7.9) with (7.10) indicates that the profit-maximising and socially optimal levels of q and a will differ. The precise relationship between them is best determined graphically. Figure 7.5 plots the equations (with the accumulation constant a on the horizontal axis and surplus s on the vertical). The comparison of social and private optimal values depends on the sign of $\alpha'(a)$, as the figure shows. If $\alpha' > 0$, then firms choose an accumulation constant which is higher than the socially optimal level. Intuitively, firms choose to increase a (relative to the social value) in order to increase the price elasticity of demand; the higher elasticity means that the firms can exercise greater monopoly power.

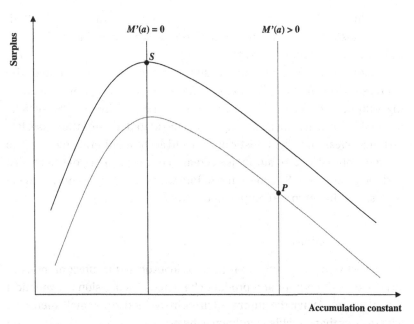

Fig. 7.5. Quantity and accumulation constant in the social optimum and oligopoly.

Notice also that when $\alpha' > 0$, surplus s in the profit-maximising equilibrium is lower than in the social optimum. It can be shown (see Dixit, 1979) that this means that output per firm is higher, but the total number of firms is lower, than the socially optimal levels. A similar analysis can be undertaken when $\alpha' < 0$: in this case, the profit-maximising equilibrium involves a lower value of the accumulation constant and of the surplus s. (Conclusions about q and n are, however, ambiguous.)

Once again, the question of whether the market output and accumulation constant will be greater than the socially optimal level boils down to an empirical question about the nature of demand (just as in the previous section; see equation (7.5)) – does the price elasticity of demand increase or decrease as the accumulation constant of the chemical rises?

Imperfect competition and externalities

The previous section showed that when the number of manufacturers of an agricultural chemical is small, the industry may produce too much of the chemical, with too high an accumulation constant; but there is also the possibility that output and the accumulation constant will be too low. Both forms of distortion lead to an unambiguous decrease in welfare. If, in addition, the social planner considers damage caused by the accumulation of the product in the environment, these conclusions must be modified. The decrease in welfare in the former case increases, clearly – profit-maximising output and accumulation increase further relative to the socially optimal levels when social damages are considered. In the latter case, it is no longer clear that oligopolistic output and accumulation are too low. The socially optimal level of output is decreased by the presence of externality, and the production distortion due to imperfect competition may actually be desirable in this more complicated setting.

Conclusions

The objective of this section has been to investigate the effect of industry structure on the output and product characteristics decisions of chemical manufacturers. When the number of firms in the industry is small, then each firm realises that it is able to influence the price it receives by its output and accumulation constant choices. If increasing the accumulation constant of their product decreases the price elasticity of demand, then firms will produce too much of chemicals with socially excessive persistence. In this case, firms reduce the amount of competition that they face (the lower the price elasticity, the lower the degree of competition) by increasing accumulation. Output decisions are also distorted, so that each firm produces too much (compared to the social optimum). These distortions unambiguously decrease welfare. The presence of externalities means that few firm conclusions can be drawn. The production distortion caused by imperfect competition may be socially desirable when chemical use gives rise to damages.

Notes

1 Financial support from the European Science Foundation, the ESRC (grant number R00429324220) and the EC (project number EV5V-CT94-0382) are gratefully acknowledged.
2 Although the discussion will be in terms of agricultural chemicals, the framework used is more general, and applies to many other chemicals.
3 'She' is used generically here for both male and female.
4 For a full economic analysis of this issue, see Beath and Katsoulacos (1991), Chapt. 5.

5 See Söderqvist Chapter 3, this volume for a quantitative interpretation of this concept.

6 For herbicides, this index might be the GUS index, which is a combination of the solubility and half-life of the chemical in water.

7 With usual assumptions placed on p and $C(Q_p)$ increasing and strictly concave, C increasing and strictly convex in both arguments), these marginal conditions are both necessary and sufficient to find the optimum.

8 Again, with the same assumptions placed on p and C as above, and assuming D is convex in both arguments, these marginal conditions are both necessary and sufficient to find the optimal solution.

9 Technically, the direction of the inequality in equation (7.5) is determined by the variation of the price elasticity of demand with the accumulation constant of the chemical (which in turn is determined by the sign of a third-order cross partial derivative of the farmers' production function).

10 To be more explicit: if the firm were to charge a slightly higher price than the one prevailing in the market, it would sell nothing – consumers would simply buy from rival firms charging a lower price. If the firm were to charge a lower price, it would gain the whole market, but would make a loss (at any quantity sold); and so it would not pursue this policy. The best that the firm can do is to charge the prevailing market price and break even. In this sense, the firm is powerless to alter the price of its product.

11 In fact, marginal revenue equals the price *minus* a (positive) term which measures the extent to which the firm is able to influence price by a change in its output (known as the elasticity of demand).

12 The analysis of this section follows that of Dixit (1979) and Spence (1977).

13 This is a verbal description of the shapes of the functions $\phi(.,.)$ and $G(.)$, which are both increasing and concave.

References

Beath, J. and Y. Katsoulacos, (1991). *The Economic Theory of Product Differentiation*, Cambridge University Press, Cambridge.

Dixit, A. (1979). Quality versus quantity competition. *Review of Economic Studies*, **46**, 587–99.

Infomark, (1993). *Market Survey*, Du Pont de Nemours and Co., Inc.

Lancaster, K. (1966). A new approach to consumer theory. *Journal of Political Economy*, **74**, 132–57.

Nadai, A. (1995). The E.U. regulatory process in pesticide registration. *CERNA Working Paper*.

Spence, M. A. (1977). Nonprice competition. *American Economic Review Papers and Proceedings*, **67**, 255–59.

Varian, H., (1992). *Microeconomic Analysis*, 3rd edition, Norton, New York.

8 Regulatory failure[1]

Robin Mason

Introduction

Chapter 7 showed that market failure (the presence of externalities and imperfect competition in product markets) may give rise to agricultural chemicals which will accumulate in the environment to a socially excessive level. It is the role of governments to intervene to construct institutions to counteract the market failure. Yet all too often, government intervention serves only to exacerbate the original problem by supplying an inappropriate or ineffective institution; such an event is termed a 'regulatory failure'. In this chapter, various sources of regulatory failure are examined.

In many cases, governments have banned particular chemicals deemed to have accumulated to excessive levels in natural resources. Rölike (1996) reported that by 1993, 78 countries had used this type of regulation. In Germany, 300 of the (approximately) 1000 pesticides registered have been banned individually. Many European countries have responded to a European Commission Directive on drinking water by banning the sale and use of particular chemicals (see p. 218). The objective of this form of regulation is to reduce chemical accumulation by two means: firstly, by removing from use chemicals already present at high concentrations; and, secondly, by providing a signal of society's disapproval of excessive accumulation. It also has the advantage of a degree of certainty (although even this is mitigated by the persistence and continued accumulation of products for several years after the imposition of a ban).

This chapter assesses the combination of maximum acceptable concentrations (MACs) and product-specific bans (PSBs) used by regulators. First, it analyses whether this approach is 'near optimal', in the sense of approximating the ideal 'first best', in which a social planner has full knowledge of all aspects of the problem, and in particular the damages associated with chemical use. It then analyses the opportunities for strategic behaviour that a MAC/PSB presents to the manufacturers of chemicals. A case study of the Italian herbicide market is presented to assess the qualitative evidence for the theories discussed.

Regulation under ignorance

This section answers the question 'is banning an individual chemical once it accumulates to a critical level in the environment a rational regulatory strategy?' The 'performance' of this regulatory approach is examined to assess how well it compares to the 'ideal' method of regulation. (All terms appearing in quotation marks will be defined in the course of the section.) The starting point of the analysis, as in Chapter 7, is to review the choices that would be made by a benevolent, omniscient and omnipotent social planner. These choices will then be compared to those made by a regulator who has only partial information and who uses MAC/PSB regulation in an attempt to compensate for this. The purpose of this comparison is to isolate the inefficiencies which arise solely from this particular approach to regulation; problems due to strategic behaviour of firms are analysed later.

The first best

The first step is to define the 'ideal' mentioned above. This will be taken as the choices of a social planner who is fully informed about all aspects of the problem. This includes the benefits derived by society from the use of agricultural chemicals, the cost of their production, and the dynamics of environmental degradation, for both present and future generations. The critical piece of information for the analysis, however, is the social damage done by chemical use, or, equivalently, the social utility derived from an uncontaminated environment. The 'first best social planner' (which will be denoted as FB) knows social damages from chemical use and incorporates them fully into her[2] choices.

The social planner maximises the discounted sum over time of social utility from chemical use.[3] Utility is defined as the total surplus (area under the inverse demand curve) minus production costs from a good (in this case, agricultural chemical). In the first best, surplus is derived from two factors – use of the chemical, and the stock of environmental quality. Hence $U^{FB} = U^{FB}(Q,S)$, where U is the social utility function, Q is total chemical use, and S is the amount of environmental quality. The inclusion of S in the utility function indicates that FB considers not only the benefits of chemical use, but also the costs in terms of damage to the environment. This damage occurs, for example, as a result of polluting groundwater reserves; in general, S will be the unpolluted stock of a natural resource.

The ignorant regulator

Why do regulators use MAC/PSBs? One possible explanation is that regulators and governments have adopted the MAC/PSB approach in an attempt to limit damages whose existence is known, but whose magnitude and probability of occurrence is not. Regulators may be aware of the possibility that contamination of natural resources could have damaging effects on human beings and the environment. Often, however, the extent and likelihood of the possible damages are unknown. This problem is particularly acute for the low toxicity, high persistence chemicals that are the focus of this chapter. The lesson has been learned from the case of DDT – any chemical can give rise to toxicological effects if it is sufficiently persistent. The long-term nature of the problem, however, makes it very difficult to foresee the harm that might occur. Regulators must act in ignorance.[4]

An extreme representation of the 'ignorant regulator's' (denoted IR) position in this case is that she does not care explicitly about the utility derived from environmental quality; after all, she has no idea how beneficial it is. Therefore, only chemical use appears in IR's utility function ($U^{IR} = U^{IR}(Q)$).[6] But she attempts to constrain future damages by imposing a MAC/PSB on chemicals. IR's problem is modelled in the following way. The objective of IR is similar to that of FB: maximise the present discounted value of an infinite flow of utility. IR realises, however, that chemical use leads to pollution, and so she regulates by banning the polluting chemical in perpetuity if it accumulates beyond the MAC.

The dynamics of environmental pollution

In order to simplify the analysis, suppose that there are only two possible chemicals that can be used. The first has a low unit production cost (without loss of generality, set equal to zero), but causes environmental damage by accumulating in a natural resource. Sales of this chemical will be denoted by x. The second chemical (known as the backstop) is more expensive: it has a constant unit cost of production of $c > 0$; but it does not accumulate in the environment. (One way to motivate this set-up is to argue that it is possible to develop pesticides of sufficiently low persistence that they would not accumulate in the environment; but these pesticides will be expensive to develop and manufacture.) Sales of this chemical will be denoted by y. This simple story captures the idea that environmental damage can be avoided (by using the second chemical) but only at some cost (the higher cost of production c).

An important distinction for the analysis is whether the resource in question is non-renewable or renewable. In the former case, pollution of the resource is irreversible – the stock of a chemical accumulated in the resource can never decrease. So, if $S(t)$ is the amount of environmental quality (the stock of unpolluted natural resource) at time t and $x(t)$ represents sales of the polluting chemical in the same time period, then the change in environmental quality can be written:[7]

$$\frac{dS(t)}{dt} = -x(t)$$

A standard example of a non-renewable resource is coal: extraction depletes the total stock irreversibly.[8] Other resources may be approximately non-renewable. For example, groundwater in geographies with non-porous rock may have recharge rates that are so low that the resource is, in effect, non-renewable. In the latter, renewable case, the resource is able to regenerate so that, in the absence of a flow of polluting chemical, the stock of unpolluted resource will grow. The change in the stock of the unpolluted resource in this case can be written as:

$$\frac{dS(t)}{dt} = -x(t) + R(S(t))$$

where $R(S(t))$ is a recharge function which describes how the resource renews itself at time t when the existing stock is $S(t)$.[9] Notice that a non-renewable resource is one whose recharge function R equals zero.

The intial stock of the unpolluted natural resource, before chemical use commences, will be denoted S_0 (i.e. the stock at time $t = 0$). MAC/PSB regulation can be interpreted as saying that the chemical is banned once the stock of the natural resource drops below a critical level $\underline{S}$.[10]

In the remainder of this section, the policies of the fully informed social planner FB and the ignorant regulator IR will be analysed and compared, first for the case of a non-renewable resource, then for a renewable resource.

Non-renewable resources

The first best (FB)

First, the choices of FB are analysed (see Krautkraemer, 1986, for a similar model). As discussed above, the objective of FB is to maximise the present discounted value of an infinite flow of utility derived from both chemical use and environmental quality, taking into account any production costs, subject to the polluting effects of chemicals. It can be shown that FB will pursue a policy with the following features:[11]

(1) Only the first, cheaper and polluting chemical is used during some initial time period, up to time T^{FB}. As a consequence, the stock of unpolluted resource decreases during this time period.
(2) The rate of use of the first chemical declines during this initial phase.
(3) At time T^{FB}, use of the polluting chemical stops and substitution to the more expensive, clean chemical occurs. Sales of the substitute are constant over time at the level $\bar{y}$ where price equals the marginal cost c. The polluting chemical is banned in perpetuity.
(4) Use of the polluting chemical stops before the stock of unpolluted resource is exhausted (i.e. $0 < S(T^{FB}) = S^{FB} < S_0$).

The intuition for these results is straightforward. During the initial period before T^{FB}, the unpolluted resource is relatively abundant and so its (marginal) value is low. In this case, the planner derives greater utility from chemical use than from preserving the stock of uncontaminated natural resource; hence the planner uses the cheaper chemical, despite the pollution that it causes. The value of the resource rises, however, as it becomes more polluted (as the stock of clean environment falls). When the value of the resource rises above a critical level, greater utility is derived from a cleaner environment than from chemical use. Therefore the planner substitutes the cleaner chemical, despite its greater expense. (The resource will not be polluted entirely due to the assumption made in note 4.)

The ignorant regulator (IR)

The objective of IR is similar to that of FB: maximise the present discounted value of an infinite flow of utility. IR, however, does not consider explicitly the utility derived from environmental quality – only chemical use appears in IR's utility function ($U^{IR} = U^{IR}(Q)$). Use of the polluting chemical is regulated by the MAC, which states that the stock of unpolluted natural resource cannot fall below the level $\underline{S}$. IR's optimal policy (also considered by Krautkraemer (1985)) can be shown to have the following features:

(1) Only the first, cheaper and polluting chemical is used during some initial time period, up to time T^{IR}. As a consequence, the stock of unpolluted resource decreases during this period.

(2) The rate of use of the first chemical declines during this initial phase.

(3) At time T^{IR}, use of the polluting chemical stops and substitution of the more expensive, clean chemical occurs. Sales of the substitute are $\bar{y}$. The polluting chemical is banned in perpetuity.

(4) Use of the polluting chemical stops when the stock of unpolluted resource hits the MAC level (i.e. $0 < S(T^{IR}) = S^{IR} = \underline{S}$).

These results and the intuition behind them are very similar to those for FB.

First best versus ignorant regulation

Both FB and IR policies involve an initial period during which use of the polluting chemical declines; both involve a switch to the clean, more expensive chemical once a critical level of the resource stock is reached. Two questions are central for the comparison of the two policies: (1) how are the times T^{FB} and T^{IR} (at which chemical use switches from the first to the second product) related?; (2) how does use of the first chemical in the IR programme compare to the first best? The answer to these questions are critical in determining how the MAC/PSB regulation both preserves environmental quality and incurs higher production costs by restricting use of the cheaper, polluting chemical.

Consider the first question. If $\underline{S}$ is small (i.e. the MAC ($S_0 - \underline{S}$) is high), then T^{IR} will be large. Intuitively, this is obvious enough – it takes longer for the polluting chemical to reach its MAC when this concentration is high. (A formal proof can be found in Stiglitz and Dasgupta, 1982, although some adaptation is required; see Mason, 1997). Next, suppose that $\underline{S} = S^{FB}$, so that both FB and IR pollute the natural resource to the same extent. Then it can be shown that the appearance of the stock of the natural resource in the utility function of FB means that the time T^{FB} is greater than T^{SB}. Again, this result is intuitive – valuing directly environmental quality should lead to a slower rate of pollution. (The formal proof, however, is quite involved; see Mason, 1997.) Combining the two results yields the following conclusion:

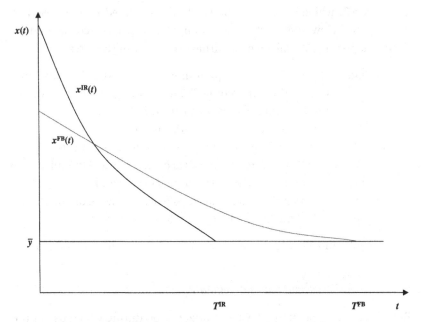

Fig. 8.1. Sales of the polluting chemical by the first best and ignorant regulator: case 1.

provided the MAC resource stock $\underline{S}$ is sufficiently large (i.e. is higher than some level $\tilde{S}$ which is less than S^{FB}), then IR will substitute to the backstop sooner than FB (i.e. $T^{IR} < T^{FB}$). Only if $\underline{S}$ is very low (less than $\tilde{S}$) will IR substitute after FB. The conclusion is summarised in the diagram below.

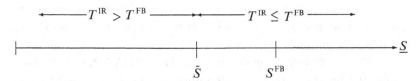

Consider now the second question: how does use of the first chemical in IR's policy compare to the first best? Again, the answer to the question depends on how the MAC level $\underline{S}$ compares to the extent of first best environmental preservation S^{FB}. Figure 8.1 plots sales of the polluting chemical $x(t)$ on the vertical axis against time on the horizontal axis for both FB and IR. (The former is labelled $x^{FB}(t)$, the latter $x^{IR}(t)$.) It shows the case where $\underline{S} = S^{FB}$, so that $T^{IR} < T^{FB}$. The area under the $x(t)$ schedule and above the line $\bar{y}$ must equal the total polluted resource stock $(S_0 - \underline{S})$. Because $T^{IR} < T^{FB}$ and $x^{IR}(T^{IR}) = x^{FB}(T^{FB})$, it must be that the sales schedules cross, as shown.[12] Indeed, while $\underline{S}$ is greater than $\tilde{S}$ (so that $T^{IR} < T^{FB}$)

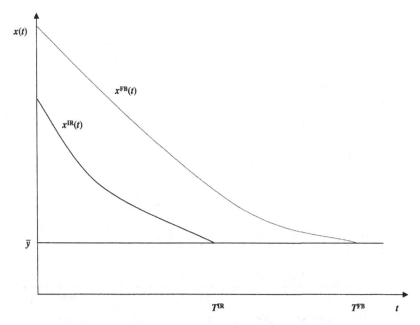

Fig. 8.2. Sales of the polluting chemical by the first best and ignorant
regulator: case 2.

and smaller than some $\check{S} > S^{FB}$, it must be the case that the output schedules of FB and IR cross, with $x^{IR}(0) > x^{FB}(0)$. Should $\underline{S}$ exceed $\check{S}$, then
output in IR's policy will lie entirely below FB's output schedule (see
Figure 8.2). Once $\underline{S}$ falls below $\check{S}$, then T^{IR} is greater than T^{FB}; then, provided $\underline{S}$ is greater than some $\hat{S}$, IR's output schedule lies entirely above
FB's schedule (Figure 8.3 shows this case). Finally, if $\underline{S}$ is very low (less
than $\hat{S}$), then, once again, the output schedules of the IR and FB policies
must cross (Figure 8.4). In this case, FB's output starts off higher but
eventually falls below output in the IR policy. The results are summarised
in the diagram below.

	$x^{IR}(0) > x^{FB}(0)$	$x^{IR}(0) \leq x^{FB}(0)$
$T^{IR} \leq T^{FB}$	$\tilde{S} \leq \underline{S} < \check{S}$	$\underline{S} \geq \check{S}$
$T^{IR} > T^{FB}$	$\hat{S} \leq \underline{S} < \tilde{S}$	$\underline{S} < \hat{S}$

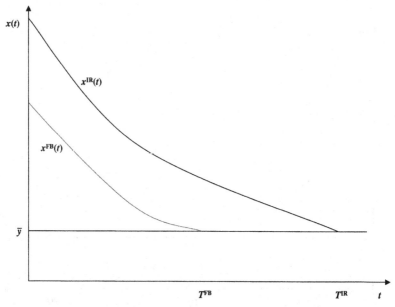

Fig. 8.3. Sales of the polluting chemical by the first best and ignorant regulator: case 3.

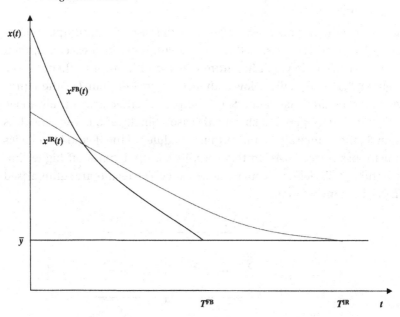

Fig. 8.4. Sales of the polluting chemical by the first best and ignorant regulator: case 4.

These results allow a precise assessment of the inefficiencies of MAC/PSB regulation. Even if the ignorant regulator is fortunate enough to set the MAC at the first best level (i.e. $\underline{S} = S^{FB}$), then IR's policy will cause excessive production costs due to early substitution to the more expensive chemical.[13] If IR is very cautious, setting a low MAC (and so a high $\underline{S}$), then production inefficiencies increase (although so does environmental preservation, of course). Not only will IR substitute the more expensive chemical earlier than FB; IR will also use inefficiently low amounts of the polluting chemical at all times before the chemical is banned. Naturally, IR might avoid these production inefficiencies by setting a high MAC (low $\underline{S}$). But the results above show that $\underline{S}$ will need to be considerably lower than S^{FB} in order to avoid the costs of early substitution ($\underline{S} < \tilde{S} < S^{FB}$). There is a clear trade-off with environment quality in this case.

Renewable resources

The discussion now turns to the case of renewable resources. It will become clear that the policy implications in this case are quite different; in particular, the desirability of product bans will come under question. A complete analysis would involve full welfare comparison of the first best and ignorant regulator's programmes. Such generality is beyond the scope of this chapter; instead, the discussion will be confined to the comparison of two particular aspects of the programmes: the impact on the level of environmental degradation, and the cost of substituting to more expensive, but less polluting chemicals. It will be shown that the MAC/PSB approach leads generally to a lower level of environmental quality than the first best, and incurs higher costs. Although these two factors do not capture total welfare in its entirety, they are important indicators of the potential inefficiencies of the second best regulation.

The first best (FB)

An important concept for the analysis is that of the *steady state*. As the name suggests, it is the point at which all quantities (chemical use and the stock of unpolluted natural resource) are constant, not changing over time. An important feature of the steady state is that it involves use of the polluting chemical; in other words, FB will not impose a permanent product ban. The reason for this is clear: when the resource is renewable, it is optimal in the steady state to use just enough of the polluting chemical to ensure that there is no net change in the stock of the unpolluted resource. So steady-state sales of the first chemical are given by:

$$x = R(S)$$

Two steady states are possible for FB; they are characterised as follows:

(1) The cheaper, polluting chemical is used, with no production of the more expensive, cleaner chemical. The steady-state levels of use of the two chemicals are $x_1^{FB} > 0$ and $y_1^{FB} = 0$. The steady-state stock of unpolluted resource is S_1^{FB}, and is greater than zero.[14]

(2) Both chemicals – the cheaper, polluting chemical and the more expensive, cleaner backstop – are used in the steady state. Steady-state use of the two chemicals is denoted x_2^{FB} and y_2^{FB}, both of which are greater than zero. The steady-state stock of the resource is $S_2^{FB} > 0$.[15]

Which steady state occurs depends on the parameters of the problem. For example, if c is small (i.e. the substitute chemical is relatively cheap to produce), then it is likely that the steady state will involve use of the backstop.

The steady state is a long-run concept; it may take a significant amount of time to reach. It is important, therefore, to understand what happens before the steady state is attained. If S_0 is sufficiently large (greater than S_1^{FB} or S_2^{FB}), then use of the cheaper polluting chemical starts at $t = 0$, and declines monotonically over time to the appropriate steady-state level. If S_0 is very low, however, the optimal path may involve an initial period during which only the non-polluting backstop is used. During this time, the resource stock increases according to the recharge function $R(S)$. Use of the polluting chemical may commence before the steady state is reached; if this is the case, its use starts at a low level, and increases monotonically over time until the steady state is reached. Alternatively, use of the polluting chemical may not occur until the resource stock reaches its steady-state level; at this time, use of the polluting chemical jumps from zero to its steady-state level. In the discussion that follows, it will be assumed that S_0 is large, so that $x > 0$ at $t = 0$.

The ignorant regulator (IR)

The analysis is conceptually very similar to that of the previous section. The renewability of the resource means that there will be some use of the polluting chemical in the steady state (i.e. no perpetual ban should be insti-tuted). Steady states will be distinguished by whether there is use of the second, non-polluting chemical or not. There is an additional complica-tion, however, introduced by the MAC regulation. This regulation prevents the stock of the unpolluted resource falling below $\underline{S}$; and so steady states will be distinguished also by whether this constraint is binding or not. Thus there are four possible steady states:

(1) The cheaper, polluting chemical is used, with no production of the more expensive, cleaner chemical. The steady-state levels of use of the two chemicals are $x_1^{IR} > 0$ and $y_1^{IR} = 0$. The steady-state stock of unpolluted resource is S_1^{IR}, and is greater than $\underline{S}$.[16]

(2) Both chemicals – the cheaper, polluting chemical and the more expensive, cleaner backstop – are used in the steady state. Steady-state use of the two chemicals is denoted x_2^{IR} and y_2^{IR}, both of which are greater than zero. The steady-state stock of the resource $S_2^{IR} > \underline{S}$. (In fact, $x_2^{IR} = x_1^{IR}$ and $S_2^{IR} = S_1^{IR}$.)[17]

(3) The cheaper, polluting chemical is used, with no production of the more expensive, cleaner chemical. The steady-state levels of use of the two chemicals are $x_3^{IR} > 0$ and $y_3^{IR} = 0$. The steady-state stock of unpolluted resource is $S_3^{IR} = \underline{S}$, i.e. the MAC constraint is binding.[18]

(4) Both chemicals – the cheaper, polluting chemical and the more expensive, cleaner backstop – are used in the steady state. Steady-state use of the two chemicals is denoted x_4^{IR} and y_4^{IR}, both of which are greater than zero. The steady-state stock of the resource is $S_4^{IR} = \underline{S}$.[19]

The paths to the steady state are as described for the FB case. A tempo-rary ban on the polluting chemical may be required if the initial stock of unpolluted resource is low; once the resource has recharged sufficiently, the polluting chemical may be used. Otherwise the polluting chemical can be used immediately, with its sales declining over the period up to the steady state. As before, it will be assumed that S_0 is sufficiently large that $x > 0$ at $t = 0$.

First best versus ignorant regulation

Both FB and IR programmes involve an initial period during which use of the polluting chemical declines (assuming S_0 is sufficiently high). Both involve a steady state in which there is some use of the polluting chemical and some preservation of environmental quality. How do the steady states of the system compare?[20] Two aspects of the steady state are of interest: (1) the stock of unpolluted resource and sales of the polluting chemical; and (2) the cost of substitution to the more expensive chemical. These are analysed in turn below.

If the MAC constraint is not binding (because the MAC is set high, i.e. $\underline{S}$ is low), then it is straightforward to show that the stock of unpolluted resource is lower in the IR programme than in the FB programme, i.e. $S^{IR} < S^{FB}$. This is intuitive enough – if IR does not protect the environment adequately by setting low MACs, resource degradation will exceed first best levels. There is an interesting consequence in this case, related to the renewability of the resource. The steady-state value of IR's resource stock will be too low; and so will its use of the polluting chemical (since $x^{IR} = R(S^{IR})$ and the recharge rate is higher for larger stocks). So the steady-state inefficiency is two-fold: too low a level of environmental preservation and too low a usage of the cheaper chemical. When the MAC constraint binds, it is clear that comparison of the steady-state values of x and S will depend on the exact level of the MAC. If $\underline{S} \leq S^{FB}$, then again it can be concluded that $S^{IR} \leq S^{FB}$ and $x^{IR} \leq x^{FB}$. If the opposite holds, then the steady state level of unpolluted resource is higher in the IR programme than in the FB programme (and so $x^{IR} > x^{FB}$). The main message of the comparison is that IR will almost certainly attain an inefficient steady state, except in the fortuitous and unlikely event that she sets the MAC at the first best level: $\underline{S} = S^{FB}$. And even in this case, the path to the steady state will be inefficient.

The cost of substitution to the more expensive chemical is now examined. Consider first the case where the MAC is high, so that $\underline{S} < S^{FB}$. Then $x^{IR} < x^{FB}$ (from the arguments above); and the marginal utility from chemical use will be higher for IR than for FB (since IR is using less of the polluting chemical and diminishing marginal utility is assumed). So IR will want higher use of chemicals; but she is constrained by the MAC not to use the polluting chemical. This argument suggests (and it can be shown formally) that it can never be the case that FB will use the clean chemical in the steady state while IR does not. It may be the case, however, that IR uses the backstop while FB does not. Therefore when $\underline{S} < S^{FB}$, the steady-state costs of chemical production in the IR programme are at least as great (and may be greater) than in the FB programme. This raises the paradox that the ignorant regulator who sets a lax MAC may raise the cost of chemical use, as well as lowering the steady-state stock of the natural resource.[21]

Summary

This section has shown that the regulatory approach of using maximum acceptable concentration limits followed by product-specific bans (MAC/PSB) may not approximate the ideal outcome that would be achieved if the social planner had full knowledge of all aspects of the problem. An immediate point to note is that permanent product bans, as practised by many governments, are optimal only if the resource in question is non-renewable. If the resource is renewable, then optimal management requires that (at the steady state) use is made of the polluting chemical at the recharge rate of the resource. Temporary bans may be required to allow the resource to replenish itself if pollution in the past has been severe. If the resource is non-renewable, then it was shown that there is a clear trade-off between environmental preservation and the cost of chemical production. In order to avoid inefficiently early substitution to the more expensive chemical and under-usage of the cheaper product, the ignorant regulator must set a very high MAC (so that $\underline{S}$ must be considerably lower than S^{FB}). The consequence will be excessive pollution of the natural resource. If the resource is renewable, then there will be inefficiency at the steady state except if the ignorant regulator is fortunate enough to have set the MAC at the first best level (there is, of course, no way that this can be guaranteed). If the MAC is set too high, then the inefficiency may involve all aspects of the problem: use of the cheaper chemical will be too low, the resource will be polluted to an excessive extent, and there will be higher production costs due to increased use of the more expensive substitute. This rather counter-intuitive outcome suggests that regulators may be better advised to set lower MACs and to allow sustainable use of polluting chemicals.

Of course, these conclusions must be qualified by noting that the MAC/PSB approach has been analysed in isolation. It is not possible to conclude from the results of this section that this form of regulation should be abandoned; before this can be said, the performance of this approach relative to other candidates must be assessed. The analysis suggests, however, that there is room for improvement. The development of alternative regulatory approaches is the subject of other chapters in this book; to conclude this section, some simple observations will be made. The outcome in the ideal, first best case depends on three factors: (1) the rate of recharge of the natural resource, (2) the discount rate r, and (3) the relative cost of the non-polluting chemical c. This fact suggests that any regulatory approach should, in contrast to that adopted in, for example, Italy, be both location and chemical specific. Instead of a nationwide ban on the use of a chemical whose concentration in drinking water has exceeded a maximum level which is the same for all chemicals, optimal regulation will require MACs which vary according to the chemicals' characteristics and use restrictions that differ depending on the recharge rate of natural resources in different locations.

Strategic behaviour and the protection of future patent value

The focus of the analysis now moves away from direct welfare comparison to consideration of the opportunities that the MAC/PSB approach can give manufacturers to behave strategically. This section analyses how a firm's behaviour alters when it attempts to protect the value of future patents that it might win, and it describes how the MAC/PSB regulation can assist the firm in achieving this objective.

The model

To address these issues, a model is developed that is very similar in structure to the one that was used in the previous section. (The mathematical details can be found in Mason and Swanson, 1997.) The components of the model are as follows.

The product market

The model runs over the life-time of two chemicals. The first has a low unit production cost (without loss of generality, set equal to zero), but accumulates in the environment. Sales of this chemical at time t are denoted $x(t)$; it is introduced to the market at $t = 0$ and sold under monopoly by the inventor until expiration of the patent, which occurs at $t = t^*$. At this time, other manufacturers are allowed to enter the market, and competition in the first chemical becomes perfect. The second chemical is more expensive: it has a constant unit cost of production of $c > 0$ with sales $y(t)$. To limit the model to one 'cycle', it is assumed that the chemical does not accumulate in the environment. The two chemicals are assumed to be perfect substitutes. The choke-off price is assumed to be infinite, so that chemical use will never fall to zero (there will always be demand for chemicals, no matter how high their price).

The natural resource

As in the last section, the first chemical accumulates in a natural resource (for example, groundwater). But (unlike in the previous section) not all of the chemical reaches the environment, so that the total amount used must be multiplied by a fraction A which lies between some lower value $\underline{A}$ and 1. This 'accumulation constant' measures the ability of the chemical to accumulate in the natural resource.[22] $\underline{A}$ is greater than zero, to reflect the fact that chemical accumulation is unavoidable unless the cost c is incurred.

Let the total stock of the resource at time t be $S(t)$, with an initial level S_0. The regulator sets a maximum acceptable concentration for the chemical so that, if it reaches this level in the resource, it is banned indefinitely. This is equivalent to saying that the stock of the natural resource must not fall below $\underline{S}$; once the effective stock $(S_0 - \underline{S})$ is exhausted, the chemical is banned. The product-specific nature of the ban means that a MAC relates to each individual chemical: once one chemical is banned, the next one to emerge on the market is assigned a 'fresh' stock of resource $(S_0 - \underline{S})$. To keep the analysis simple and to highlight the strategic behaviour involved, it is assumed that the resource is non-renewable and there is no uncertainty. The dynamics of the natural resource stock can be written as follows (where t denotes time):

$$\frac{dS}{dt} = -Ax(t)$$

that is, the rate of change in the stock of the resource over time is equal to (minus) the level of pollution, which is use of the polluting chemical $x(t)$ multiplied by the accumulation constant A.

Product innovation

The timing of invention is not of direct interest – that is, this is not a model of a patent race. The simplest possible scenario has the second chemical being invented and innovated at a known fixed time τ, with the initial patent holder being the winner of the patent race for the second chemical. (See Gilbert and Newbery, 1982, for a model of patent races in which this outcome holds.) To close the model, it is assumed that the patent on the second chemical has an infinite life.

Before developing the model further, it is worth commenting on four main assumptions: (1) the two chemicals are perfect substitutes in demand, (2) competition in the first chemical is perfect after expiration of the patent, (3) the second patent is held in perpetuity, and (4) the date of innovation of the second chemical is fixed. These assumptions can be motivated, in part, by analytical simplicity; in particular, nothing important hangs on the third and fourth. Moreover, they may be reasonable descriptions of current regulations in chemical industries. Consider the use of product-specific bans and the agrochemical sector (facts about which are drawn from Nadai, 1995). Assumption (1) – perfect substitution – is reasonable given that the product specificity of bans allows replacement of a banned chemical with a product which is virtually identical.[23] Perfect competition after patent expiration (assumption (2)) can result from current regulation which releases the full dossier of data for any out-of-patent chemical to any firm that can manufacture a 'similar' product.[24] There were approximately 90 agrochemical firms engaged in the manufacture of previously patented chemicals in the EU in 1994.

The effect of strategic behaviour

How will the initial patent holder choose the production levels (output and the accumulation constant) of the first chemical? To answer this, a related question must first be answered – at what time T should the resource stock $S_0 - \underline{S}$ be exhausted? The factors at work are clearest when τ is greater than t^*, so that the backstop is not available until some time after the patent on the first product has expired. At t^*, rivals enter the market and the initial patent holder, who had previously enjoyed a monopoly in the first product, now faces perfect competition. Since the first product and the backstop are perfect substitutes, the situation does not necessarily improve for the patent holder when the second product becomes available. In order to maximise profits from the sale of the backstop, the initial patent holder will want sales of the first product to fall to zero as soon as possible after τ. This can be achieved by ensuring that the first chemical reaches the MAC and is banned. The patent holder faces, therefore, a trade-off – increasing sales or the accumulation constant of the first chemical during its patented period means that the MAC will be reached more quickly (reducing future competition); but this may reduce profits (for example, by driving down the price) over the first t^* years of sales. The patent holder's optimal production of the polluting chemical balances the loss of profits in the first t^* years with the profits gains once the backstop is on the market.

The analysis will concentrate on two questions: (1) when might the imposition of MAC/PSB regulation *increase* accumulation, rather than decrease it as a regulator might expect? and (2) how will the initial patent holder design the first chemical to protect the value of the future patent on the second chemical?

In answering the first question, the choice of accumulation constant will be ignored. Differences in the choice of A made by the unregulated and the regulated patent holder will be considered in the second question. Any change in the rate of accumulation of the polluting chemical due to regulation reflects, therefore, a change in sales of the chemical. It can be shown (see Mason and Swanson, 1997) that MAC regulation will increase the rate of accumulation of chemicals in the environment when:

(1) the time to innovation τ is short;
(2) the MAC is high;
(3) the relative cost c of the backstop is low;
(4) demand for the chemicals is very elastic with respect to price; and
(5) the discount rate is high.

All of these conditions ensure that, when the industry is regulated, there is an initial period when the price of the first chemical is lower than it is when the industry is not regulated. As a consequence, demand for the chemical is higher. This leads to increased accumulation of the chemical in the natural resource. Whether these conditions hold clearly depends on the industry and chemicals in question. The lesson from this result is not that MAC regulation will always increase the accumulation rate of chemicals. Rather, it is that regulators must assess carefully the particulars of the situation before applying MACs and PSBs.

Now consider the question of chemial design: what incentives are present for the manufacturer's choice of the accumulation constant A of the chemical? First, note that the socially optimal value of the accumulation constant is $\underline{A}$; this preserves as far as possible environmental quality and avoids the need of substitution to the more expensive backstop.[25] In contrast, in the absence of regulation, the patent holder will be indifferent to the value of the accumulation constant.[26] But the regulated patent holder's optimal choice of the accumulation constant is 1. This choice allows the patent holder to gain maximum advantage from the MAC registration. Raising A means that the MAC will be hit more quickly; this means that rivals can be excluded from the market and hence profits from sales of the second chemical protected.

There is therefore a sharp contrast between the three cases. The social planner will design chemicals which accumulate in the environment as little as possible. The unregulated firm has no particular incentives to design certain types of chemials. But a firm, subject to MAC regulation, who holds one patent and expects to win another in the future, will design chemicals that accumulate to the maximum extent in natural resources.

Conclusions

This section has used a simple model to highlight the incentives that are created by the interaction of patents and product-specific bans. Bans, by sending out a clear message as to which chemicals are socially acceptable and which are not, seem a sensible way for regulators to attack the problem of excessive accumulation of chemicals in the environment. This argument ignores, however, the interactions that a single piece of regulation may have with existing institutions. The interaction between product-specific bans and patents has been analysed. The former are designed to reduce accumulation; the latter are required to encourage innovation; the combination of the two leads to the perverse incentives that manufacturers will ensure that chemicals accumulate to banned levels. Rather than preventing the use of socially undesirable chemicals, regulators succeed only in increasing chemical accumulation further.

Case study: the Italian herbicide industry

In this section, the empirical support for the predictions of the previous section is discussed. The model of strategic behaviour by chemical manufacturers suggests that production will be distorted as firms attempt to protect the value of future patents. The distortion involves an increase in output (or sales) and the accumulation characteristics of the chemicals sold. Sales data are not available over a sufficiently long period to allow adequate testing of the first half of this prediction (manufacturers do not reveal this information, in order to keep market shares secret). There is qualitative data, however, to address the second half of the prediction.

In July 1980, the European Commission issued a Directive (80/778/EEC), setting a MAC for individual pesticides in drinking water of 0.1 μg/l, and a 'cocktail' standard of 0.5 μg/l for the total concentration of all pesticides. States failing to meet the conditions of the Directive risk condemnation by the European Court of Justice, and imposition of penalty payments (see Faure, 1994). For example, in Italy, concentrations of atrazine, a herbicide used in maize cultivation, reached this level in 1984; in 1989, 11 wells in the Veneto region recorded atrazine levels of over 1 μg/l (see Zanin et al., 1991). Local restrictions in 1986 against contaminated drinking water supplies had little effect, and a nationwide ban on the sale and use of atrazine was imposed in 1990.[27]

The purpose of the Italian ban on atrazine was to remove from use a chemical that had already accumulated to unacceptable levels in the environment; and also to provide incentives to manufacturers to produce chemicals with lower accumulative abilities. Figure 8.5 shows GUS indices for 14 herbicides. The GUS index indicates the ability of an agricultural chemical to accumulate in groundwater, and has two components. The water solubility of a chemical is measured by its partition coefficient K_{oc}.[28] The higher the coefficient, the less soluble is the chemical in water. Degradability of the chemical is measured by the (logarithm of the) half-life in soil $(t_{1/2})$. The GUS index is then defined by GUS = $\log t_{1/2}(4 - \log K_{oc})$; herbicides are classified as water leachers (GUS > 2.8), non-leachers (GUS < 1.8), or transitional ($1.8 \leq$ GUS ≤ 2.8).

Chemicals labelled with numbers are herbicides used in Italy before the banning of atrazine (labelled with '1' in the figure). Immediately striking is the clustering of all but three of these chemicals at moderate solubilities and half-lives, and hence with transitional to leaching values of the GUS index. The chemicals labelled with letters (a, b, c) are three herbicides that have been registered for use in Italy after the imposition of the ban on atrazine. The GUS indices of two of these products are greater than those of the previous substitutes for atrazine (linuron and terbutylazine, labelled '7' and '11', respectively); the GUS index of the third is comparable to previous levels. Figure 8.5 shows clearly, therefore, that the atrazine ban has not

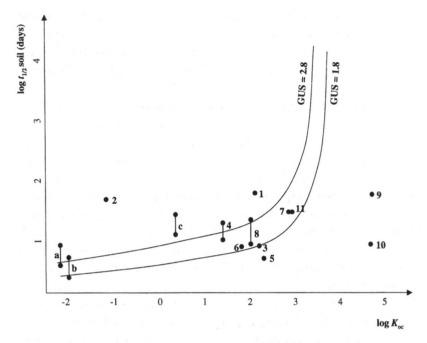

Fig. 8.5. GUS indices of herbicides. (From Vighi and Zanin, 1994.)

succeeded in decreasing the GUS indices of herbicides used in Italy; if anything, the accumulation ability of chemicals has increased since 1990.

This is obviously not a conclusive test of the predictions of the previous
section on strategic behaviour. The evidence does not suggest, however,
that firms are not reacting in a simple fashion to the regulation introduced
in Italy – the ban on atrazine has not achieved the obvious objective of a
shift in chemical characteristics. The theory given on pp. 214–18 offers
some explanation for the subsequent increase in the GUS indices of herbicides in Italy.

Conclusions

The purpose of this chapter has been to assess from a theoretical stand-
point whether the currently popular method of regulating chemical use in
Europe – banning specific chemicals once they reach a maximum acceptable concentration – is effective in reducing the accumulation of chemicals
in the environment. The conclusion that has been reached is a qualified

'no'. The second section compared this method of regulation with an 'ideal', and showed that it falls short. Permanent product bans are called for only when the resource is non-renewable; otherwise, temporary bans may be necessary, but optimal management of resource calls for sustainable (positive) use of polluting chemicals. Even when the resource is non-renewable (so that bans are justified, in a certain sense), MAC/PSB regulation may lead to costly substitution of more expensive chemicals when the ideal calls for managed use of cheaper (more polluting) chemicals. In those cases where MAC/PSB does not lead to excessive production costs, it causes too little preservation of environmental resources and too little use of chemicals. If the resource is renewable, then MAC/PSB regulation causes long-run inefficiencies except if the regulator is fortunate enough to have set the MAC at the first best level (there is, of course, no way that this can be guaranteed).

The second section considered only the choices of a 'social planner' – a benevolent regulator concerned only with the interests of society. The third section examined the opportunities for strategic behaviour created by the MAC/PSB regulation. It was shown that manufacturers will ensure that chemicals accumulate to their MACs, and hence are banned, in order to limit competition between existing products and future products on which they have patents. This strategy requires firms to increase both the sales and accumulation abilities of chemicals.

Despite the unequivocal predictions that have been made in this chapter, the welfare conclusions have been ambiguous. This has been for two reasons. First, it is not possible to conclude from any of the inefficiencies that have been identified that the MAC/PSB approach should be abandoned in favour of another – other forms of regulation must be assessed similarly before such a statement can be made. (This is done in other chapters in this book.) Secondly, the welfare implications of the strategic effects that have been identified are complex and subtle, and cannot be assessed without detailed information on market structure, social preferences, and so on. The analysis has indicated, however, some of the issues that must be considered when designing and implementing regulation – it has shown that regulation should be tailored to the characteristics of individual chemicals, and to the location of their use; it has indicated also that any proposal must be assessed carefully for the perverse incentives that regulatory intervention can all too easily create.

Notes

1 Financial support from the European Science Foundation, the ESRC (grant no. R00429324220) and the EC (project no. EV5V-CT94-0382) is gratefully acknowledged.

2 She is used generically for male and female throughout this chapter.

3 The model developed in this section is based on that of Stiglitz and Dasgupta (1982) and Dasgupta *et al.* (1982).

4 U^{FB} must satisfy a set of technical conditions which will not be considered in detail here. The most important is that the marginal valuation of both chemical consumption and the natural resource goes to infinity as consumption/the resource stock go to zero. This means that FB will never allow either chemical use or the natural resource stock to fall to zero.

5 What evidence is there that regulators act in ignorance? The Directive on pesticides in drinking water set a MAC of $0.1\mu g/l$ (see p. 219), despite scientific evidence that the safe concentration for many chemicals was at least 2 $\mu g/l$, and perhaps 20 $\mu g/l$. The limit was based on the lowest scientifically detectable limit at the time that the Directive was passed. The suggestion is that the regulators acted in anticipation of some large damage, the nature and occurrence of which was unknown when the regulation was put in place.

6 U^{IR} has technical properties equivalent to those of U^{FB}.

7 In this section, the propensity of the chemical to accumulate in natural resources (called the 'accumulation constant' in Chapter 7) is ignored. The analysis will concentrate on the sales of chemicals and demand for chemical characteristics is not considered. This simplifies the analysis considerably. Also, it is assumed that there is no lag between use of the chemical and its appearance in the natural resource. This is clearly a simplification of the more complicated dynamic process in which, for example, agricultural chemicals are transported to groundwater.

8 Ignoring discoveries of new deposits.

9 Again, technical conditions must be satisfied by the function R. It will be assumed that the necessary conditions hold; they will not be discussed in this chapter.

10 If the resource were non-renewable ($R(S) = 0$ for all values of S), then this would be entirely correct; but, provided $R(S)$ is small (which is a reasonable assumption for many natural resources, including groundwater), it will be approximately correct. Talking in terms of the stock of the resource rather than the accumulated stock of pollutant makes comparison of the various cases easier.

11 The problem can be analysed mathematically, but the treatment here will be purely discursive; interested readers are referred to Mason (1997).

12 If they do not, then the total polluted stock is less under IR than under FB; this is a contradiction, by hypothesis.

13 In this case, there is an initial period in which IR uses more of the polluting chemical than is dictated by FB's policy; but eventually use in IR's policy falls below FB use.

14 Steady-state levels of production and the resource stock are given by:

$$x_1^{FB} = R(S_1^{FB}) \tag{1}$$

$$R'(S_1^{FB}) = r - \frac{U_2^{FB}(x_1^{FB}, S_1^{FB})}{U_1^{FB}(x_1^{FB}, S_1^{FB})} \tag{2}$$

$$y_1^{FB} = 0 \tag{3}$$

(Subscripts on the function U denote partial derivatives; so for example, $U_1^{FB}(Q, S) \equiv \partial U^{FB}(Q, S)/\partial Q$.) This steady state will occur if $U_1^{FB}(x_1^{FB}, S_1^{FB}) \le c$. Equation (1) follows immediately from the definition of a steady state: in order for the resource stock to be stationary (i.e. $dS/dt = 0$), the flow of chemical pollution (i.e. x) must equal the rate of recharge of the resource. Equation (2) equates the marginal benefit from consuming one unit less of the polluting chemical (the marginal increase in the resource stock plus the marginal rate of substitution between utility from environmental quality and utility from chemical use) to the marginal cost of foregone consumption (the discount rate r).

15 Steady-state values in this case are given by:

$$x_2^{FB} = R(S_2^{FB})$$

$$R'(S_2^{FB}) = r - \frac{U_2^{FB}(x_2^{FB} + y_2^{FB}, S_2^{FB})}{c}$$

$$U_1^{FB}(x_2^{FB} + y_2^{FB}, S_2^{FB}) = c$$

This case will occur if $U_1^{FB}(x_1^{FB}, S_1^{FB}) > c$. The existence, uniqueness and stability of both steady states can be shown (see Mason, 1997).

16 Steady-state levels are given by:

$$x_1^{IR} = R(S_1^{IR})$$
$$R'(S_1^{IR}) = r$$
$$y_1^{IR} = 0$$

where r is the social rate of discount. This steady state occurs if $S_1^{IR} > \underline{S}$ and $U^{IR'}(x_1^{IR}) \le c$.

17 Steady-state values in this case are given by:

$$x_2^{IR} = R(S_2^{IR})$$
$$R'(S_2^{IR}) = r$$
$$U^{IR'}(x_2^{IR} + y_2^{IR}) = c$$

This steady state occurs if $S_2^{IR} > \underline{S}$ and $U^{IR'}(x_1^{IR}) > c$.

18 Steady-state levels are given by:

$$S_3^{IR} = \underline{S}$$
$$x_3^{IR} = R(\underline{S})$$
$$y_3^{IR} = 0$$

This steady state occurs if $S_1^{IR} < \underline{S}$ and $U^{IR'}(x_3^{IR}) \le c$.

19 Steady-state values in this case are given by:

$$S_4^{IR} = \underline{S}$$
$$x_4^{IR} = R(\underline{S})$$
$$U^{IR\prime}(x_4^{IR} + y_4^{IR}) = c$$

This steady state occurs if $S_1^{IR} > \underline{S}$ and $U^{IR\prime}(x_4^{IR}) > c$. Existence and stability of all the steady states can be shown.

20 To repeat the warning at the beginning of this section: ideally, the comparison should consider the entire time paths of the programmes. This analysis is too complex for this chapter, which will only compare the steady states of the two programmes.

21 It is clear that the converse holds. Again, it is only when IR sets the MAC so that $\underline{S} = S^{FB}$ that the steady-state production costs for IR and FB are identical.

22 To keep the analysis tractable, it is assumed that there is no demand for the characteristics of the chemical.

23 For example, the ban in Italy of the herbicide atrazine has lead to substitution to terbutylazine, a chemical from the same class as atrazine, and with very similar physical chemistry characteristics. See p. 219.

24 There is, of course, some controversy about what constitutes a 'similar' product.

25 The costs of choosing A have not been considered. If designing chemicals with low A is costly (because of the increased research and development required to find a suitable molecule, for example), then the socially optimal value of A might be greater than $\underline{A}$. It will still, however, be low.

26 There is no demand for accumulation, nor will the manufacturer be penalised in any way for accumulation of the chemical. The level of A is not relevant, therefore, for the manufacturer when maximising profits.

27 Atrazine was not the only chemical to be banned – the sale and use of alachlor on soya was also prohibited. In addition, maximum permissible doses of several chemicals were reduced significantly (see Zanin *et al.*, 1991).

28 K_{oc} measures the tendency of a substance to bind to soil solids, and is obtained conventionally by a calculation based on the K_{ow}, which is the partition coefficient for non-ionic molecules in an organic medium (octanol) versus water.

References

Dasgupta, P., Gilbert, R. J. and Stiglitz, J. E. (1982). Invention and innovation under alternative market structures: the case of natural resources. *Review of Economic Studies*, **49**, 567–82.

Faure, M. (1994). The E.C. Directive on drinking water: institutional aspects. In *Environmental Toxicology, Economics and Institutions, The Atrazine Case Study*, ed. L. Bergman and D. M. Pugh, pp. 39–87. Kluwer Academic Publishers, Dordrecht.

Gilbert, R. J. and Newbery, D. M. G. (1982). Preemptive patenting and the persistence of monopoly. *American Economic Review*, **72**, 514–26.

Krautkraemer, J. A. (1985). Optimal growth, resource amenities, and the preservation of natural environments. *Review of Economic Studies*, **52**, 153–70.

Krautkraemer, J. A. (1986). Optimal depletion with resource amenities and a backstop technology. *Resources and Energy*, **8**, 133–49.

Mason, R. A. (1997). The dynamic inefficiency of product bans. Nuffield College (Mimeo)

Mason, R. A. and Swanson, T. M. (1997). Entry deferrence and environmental regulation. *Nuffield College Economics Discussion Paper*, no. **132**.

Nadia, A. (1995). The EU regulatory process in pesticide registration. *CERNA Working Paper.*

Rölike, A. (1996). The ban of a single pesticide. *Geneva Papers on Risk and Insurance*, **21**, 224–39.

Stiglitz, J. E. and Dasgupta, P. (1982). Market structure and resource depletion: a contribution to the theory of intertemporal monopolistic competition. *Journal of Economic Theory*, **28**, 128–64.

Vighi, M. and Zanin, G. (1994). Agronomic and ecotoxicological aspects of herbicide contamination of groundwater in Italy. *Environmental Toxicology, Economics and Institutions, The Atrazine Case Study*, ed. L. Bergman and D. M. Pugh, pp. 111–39. Kluwer Academic Publishers, Dordrecht.

Zanin, G., Borin, M., Altissimo, L. and Calamari, D. (1991). Simulazione della contaminazione da erbicidi nell'acquifero a Nord di Vicenza (Italia nord-orientale). Paper presented at the 25th Annual Meeting of SIA, Bologna.

Part IV

Policies for regulating chemical accumulation

9 Optimal policies for regulating persistent chemicals

Timothy Swanson

Introduction

Some persistence is a desirable characteristic of useful chemical products, since otherwise the chemical would degrade instantaneously into inert components of little usefulness. The objective of any purchaser of a chemical product is to acquire the activity of the product. The more persistent the product, the less often the purchaser must expend the labour and capital required to apply it. For these reasons chemical persistence is not an unmitigated 'bad'; it is in fact an in-built characteristic driven by the demands of consumer groups.

However, the socially optimal degree of persistence is not necessarily ensured through market mechanisms, since persistence redounds to the benefit or detriment of many individuals other than the purchaser. Specifically, since chemicals that are persistent must be active within one of the various basic environmental media (atmospheric, hydrological or organic), this activity will also be experienced by the many others in contact with the same medium. Since this activity may be undesirable from the perspective of the many other persons subjected to it, they might prefer a lower level of persistence than would the individual who is making the purchase decision without taking their preferences into account.

It is the internalisation of this externality that is the subject of this chapter. Policies must be designed in order to mitigate the problem of greater-than-optimal product durability when persistence ensures that people other than the consumer will feel the impacts of the product. The particular policies that we investigate here are those that will cause manufacturers and users to take the correct decisions in the design and application of agricultural pesticides. This chapter proceeds by initially setting forth the range of information that is relevant and available for the construction of these policies, and then discussing the various parameters that are important to take into consideration when developing optimal policies.

Agricultural chemical characteristics: their design and their measurement

Chemicals are useful because of three basic characteristics: affinity, persistence and (in agriculture) toxicity. Affinity determines the environmental compartment or medium through which the chemical substance will travel during its active life. There are three compartments into which chemicals partition themselves: organic (living matter or biota), hydrological (water) or atmospheric (air). Depending upon the application for which the substance is being used, the chemical is usually designed to operate in only one. For example, an insecticide will usually be designed to channel itself through the organic compartment in order to react with living pests. A herbicide will usually be designed to channel itself through the hydrological compartment, using the flow of water as a vector to inflict itself upon the seeds of vegetative matter in the soil. An affinity for a particular environmental compartment means that the chemical substance is much more likely to react with and within the designated compartment than with any of the others, and the characteristic of affinity is in-built by manufacturers in order to direct the chemical toward the substances with which it should react.

The characteristic of persistence concerns the rate at which the chemical will react within the designated medium. If a chemical has a very high rate of reactivity, then it will react with substances within the environmental compartment very rapidly, thereby producing by-products of those reactions and rapidly reducing the amounts of the original chemical for future reactivity; this is known as biodegradation. A chemical has experienced complete biodegradation when the only remaining by-products are the relatively inert substances carbon dioxide and water; this is known as mineralisation. Persistence is therefore the name given to the trait of low-level reactivity. Persistence is a useful trait because it reduces the number of times that a chemical substance must be placed within an environmental medium over a given period in order to maintain its activity within that medium.

The object of agricultural chemistry is to place chemical substances in environmental media that will create opportunities for reactions with the targeted organisms. Affinity is the characteristic that is given to the chemical in order to create the opportunities for reactions with the targets. Persistence is the characteristic that is given to the chemical in order to minimise the number of applications required to retain activity within that target. Toxicity is the desired reaction between the chemical and the target when they meet.

What sort of information is available for the measurement of these characteristics? The information regarding assessment of the risks from chemical toxicity has been reviewed elsewhere (Vighi, *et al.*, Chapter 4, this volume). In this analysis we are going to assume that the health and environment costs implicit in the accumulation of a chemical substance within an environment medium have been estimated. (Press and Söderqvist, Chapter 6, this volume) The remainder of this section summarises the nature of the information available regarding the other two important characteristics determining the impact of chemicals: affinity and persistence.

Affinity and persistence together determine the environmental fate of the chemical, i.e. they determine the environmental compartment in which the chemical will ultimately reside. The information takes two basic forms: *decay rates* and *affinity quotients* (see the appendix to this chapter). Decay rates measure the rate at which the chemicals react within designated environmental media, measured in terms of the amount of time required for the original mass of chemical to be reduced by half (half-life). Affinity quotients measure the relative rate of affinity of the chemical for various types of compounds, e.g. organics (oils) versus waters. Together these measures help to predict the environmental fate of the chemical compound.

Again, it is important to emphasise that the environmental fate of a chemical substance is a choice variable for the chemical producer. All chemical substances could be made to decay instantaneously, and thus not accumulate, but this would imply zero effectiveness. The problem is that the cost of in-built persistence is implied accumulation. The studies on affinity indicate the route of accumulation that such persistence will take. The organochlorines (such as DDT) are fat soluble and hence accumulated within the food chain. Substances such as atrazine are more water soluble and hence are more problematic with respect to groundwater.

In short, pesticide manufacturers are devising chemical substances with an in-built degree of water (and fat) solubility and rate of persistence, so that rainwater will take the chemical into the ground where it will then have its desired affect on 'pests' during the period of its persistence. More generally, herbicides need to be water soluble to be absorbed by targeted plant life; insecticides need to be fat soluble to be absorbed by targeted animal life. Hence, the properties of accumulating chemicals are in-built by manufacturers to serve the purposes of their intended customers. The environmental problem is that manufacturers do not take the cost of consequent environmental contamination into consideration when making this decision. The product is being made more durable than is socially optimal, in deference to the durability requirements of the producer's market. The job of the regulator is to cause manufacturers to consider their implicit use of environmental resources when making these pesticide design decisions.

Policy relevant issues

In the remainder of this chapter I will examine how the manufacturers and users of chemical products might be caused to take into consideration the externalities resulting from the choice of certain chemical characteristics for use in agriculture. Three distinct policy issues will now be considered in turn in each of the following sections.

(1) How to determine the correct relative incentives to give to manufacturers with regard to the choice of the characteristic of persistence.

(2) How to give the correct relative incentives to manufacturers with regard to the choice of the characteristic of affinity.

(3) How to take into consideration the variety of local conditions that contribute to the rate at which a persisting chemical will accumulate within the environment. For example, how is it possible to take into account the geological/hydrological conditions under which the substance will be used if groundwater contamination is the concern? Should there be a separate set of instruments to regulate users as well as those which regulate producers?

General persistence taxes

In this section we investigate *inter alia* the general nature of the instrument that is required to regulate persistence in agricultural chemicals. We find that there will be substitution between quantities and qualities if regulation is based on either alone: hence, it is necessary to base an effective instrument on the combined product of quantities and persistence qualities. Therefore, this analysis demonstrates the need for a policy based on a per unit accumulation tax that is levied in respect to the relative persistence of the chemical substance. This is demonstrated straightforwardly in the following simple model. The analysis that follows is standard, and flows from a number of similar models (Cropper, 1976; Baumol and Oates, 1979; Conrad and Olson, 1992).

Unregulated competitive firm

Consider a competitive industry of identical firms marketing a chemical that maximises the users' preferences for affinity, persistence and toxicity, but without concern for the impacts of these characteristics on others. The firms are currently unregulated, and do not anticipate the possibility of any regulation being introduced. This scenario produces the simplest possible decision-making criterion.

Firm objective (competitive industry)

$$\max_{\{q,A\}} p(A)q - c(A)q - I_1 q \qquad (9.1)$$

where: $p(A)$ is the price of the chemical as a function of the choice of the characteristic of persistence; q is the quantity of the chemical; $c(A)$ is the cost of the chemical as a function of the choice of the characteristic of persistence; and I_1 is an instrument available to the regulator to control the industry (discussed below)

Firm maximisation conditions

$$q^*: p - c - I_1 = 0 \qquad (9.2)$$
$$A^*: p_A = c_A$$

These first-order conditions describe a point of profit maximisation for these firms so long as the described objective function is convex in the respective arguments. This is clearly the case with respect to the choice of the quantity of the chemical sold, but it is less obviously the case with respect to the choice of the persistence of the chemical. Assuming for the moment that $p'(A) < 0$ and $c'(A) > 0$, convexity is assured and there exists a unique optimum. These conditions state that the firm will produce persistence to such an extent that the marginal consumer is willing to pay for the additional cost of the last unit of persistence (assuming that it is more expensive to generate greater persistence). There is no consideration given to any person's preferences regarding persistence other than those of this one marginal consumer of the chemical product.

Regulating the production of persistence

In regulating this industry, the regulator will take into consideration the impacts on others of the choice of persistence for a given chemical product. Other persons do not care about the choice of persistence per se but they do care about how this choice will impact upon the quality of other environmental resources which they use.

Initially I will concentrate on a chemical with a given characteristic of affinity, so that the environmental compartment of concern is known *a priori*. For example, many synthetic chemicals have an affinity for water (so that this medium can be used as a vector of transport to their objectives) and the fate of these chemicals must lie ultimately within this environmental compartment. For the purposes of this section, it is assumed that the trait of affinity is a given.

The first step involves making a connection between the choice of persistence and the use of the environmental resource. Assume that it is known that the application of a particular chemical substance will on average reach the level of groundwater in a given period of time, and that the degradation process will halt within that medium (because it is out of contact with other substances, light and air.) Then the choice of a given persistence characteristic will determine the proportion of the original quantity that will ultimately accumulate within the groundwater resource. From this point onwards I will assume that the persistence characteristic is represented by a variable (A) such that $0 < A < 1$, where A represents the proportion of a quantity q that will accumulate in groundwater under average geological conditions. This creates the link between the choice of persistence (A) and groundwater quality (S); increased persistence constitutes direct use of the groundwater resource.

The second step required in the development of the regulator's model is the incorporation of other persons' preferences regarding groundwater quality. This is accomplished by including their preferences over groundwater quality ($D(S)$) directly within the social objective function. Also included within that objective function are the profits of the competitive industry producing the chemical.

The final step in the formulation of the societal object regarding the choice of persistence is the explicit incorporation of time within the problem. It is assumed that the regulator is concerned about the resource over the entirety of the time horizon, and that he or she does not discount for the lag between choice of persistence and accumulation with the groundwater. In effect, the regulator knows that the production of a chosen level and quantity of a persistent chemical product implies the irreversible accumulation of a given proportion within the groundwater resource. We denote the present value of this future accumulation as $D(S)$. This could be interpreted in two ways. First, it could be given the standard interpretation of a future impact discounted back to a current value. Alternatively, it could be argued that pristine groundwater quality is a unique resource and that its contamination is an irreversible process, in which case appeal to Krutilla and Fisher (1978) may be made to justify a different or possibly zero rate of discount applied to these future preferences. In either event we include future consumers' preferences regarding groundwater quality explicitly within the current societal objective function (i.e. we do not explicitly provide for a lag between chemical use and groundwater use) and the reader may choose which interpretation to apply to $D(S)$. All of the variables in the societal objective function should be subscripted for time, but these subscripts are suppressed here for simplicity.

Societal object regarding choice of persistence

$$\max_{\{q,A\}} \int_0^\infty e^{-rt} [D(S) + p(A)q - c(A)q]\,\mathrm{d}t$$

$$\text{subject: } \frac{\partial S}{\partial t} = -Aq + R(S) \tag{9.3}$$

$$\text{given } R' > 0,\ R'' < 0,\ S(0)$$

where $R(S)$ is the natural rate of recharge of the groundwater quality S; $D(S)$ is the societal preference for groundwater quality; and A is the persistence characteristic expressed in terms of the anticipated proportion of ultimate accumulation within the groundwater resource.

Societal maximisation conditions

Assuming the requisite concavity necessary for applying the maximum principle, we obtain the following first-order conditions (two static, one balancing) for the description of the optimal control path regarding the choice of persistence and the use of groundwater.

Static conditions

$$A^*: p' = c' + \mu \qquad (9.4)$$

$$q^*: p - c = A\mu$$

where μ is the co-state variable relating to the dynamic constraint, i.e. it is the value of an additional unit of groundwater quality for the relaxation of that constraint.

Portfolio balancing condition

$$S^*: \frac{\dot{\mu}}{\mu} - \frac{D'(S)}{\mu} + R'(S) = r \qquad (9.5)$$

$$\dot{\mu} = d\mu/dt$$

For the time being we will focus only on the differences between equations (9.2) and (9.4) and their implications for optimal policy regarding the choice of persistence. The difference in each of the two sets of first-order conditions lies in the regulator's consideration of the value of the stock of groundwater when selecting both the level of persistence and the quantity of the persisting chemical to produce. The regulator will ascribe a *rental value* (μ) to the stock of groundwater quality, and will consider the loss of this value as one of the costs implied in the increased production of persisting chemicals. In the choice of the socially optimal persistence characteristic, the regulator will consider both the increased profits to producers ($p' - c'$) but also the increased costs by virtue of the increased use of groundwater quality (μ); only when the latter exceeds the former will the regulator allow increased persistence to be selected.

The nature of the instrument required to correct for this externality is also implicit in the comparison of conditions (9.2) and (9.4). In order for the incentives of the chemical firm to be aligned with those of the regulator, all that is required is that the instrument I_1 be defined so that it equates the two. This implies the following instrument, which I have termed an 'accumulation tax'.

The accumulation tax

$$I_1 = A\mu \tag{9.6}$$

$$\text{where } \mu^{ss} = \frac{D'}{r - R'}$$

The accumulation tax charges a unit production tax corresponding to the product of: (1) the anticipated amount of ultimate accumulation in groundwater (corresponding to the proportional persistence rate of the chemical, A) and (2) the stock value of the groundwater resource (μ). This construction of the optimal instrument is based upon the theory that it is known in advance approximately how much of a given produced chemical (with persistence A in quantity q) will ultimately reside in a given environmental compartment (here, groundwater). Then it is possible to tax chemical persistence (defined here as the proportional accumulation rate) in order to cause the manufacturer to internalise use of groundwater quality when choosing the persistence of a chemical.

The actual tax rate is then based upon the value of one unit of the stock of groundwater quality, on the theory that the chemical producer may use any given unit of groundwater only if the firm is able and willing to pay the opportunity cost of that unit. What is this opportunity cost? The second equation in (9.6) gives the steady state value of the co-state variable for this system (the transition phase is discussed below), and this equation defines the precise measure of the opportunity cost of lost groundwater quality. It states that the stock value of the resource is equal to the present value of a stream of foregone water quality benefits (D') discounted by the resources 'own discount rate' ($r - R'$) into perpetuity. This means that the loss of one unit of water quality implies a loss not only to those who first experience it, but also to all future generations whose groundwater quality stock starts from a lower level than would otherwise be the case. The optimal tax will charge current users for the costs that they impose upon all other users, current and future.

It is important to note that we have defined this problem in a manner that makes it feasible to accomplish the overall objective (i.e. regulating the aggregate production of persistence) with a single instrument. This is attributable to our definition of A as the proportion of groundwater quality used per unit of chemical production. So long as this is a reasonably accurate depiction of the link between chemical use and groundwater contamination, then a single instrument is all that is required, as it is the product of the two (Aq) that must be regulated.

For example, the general characteristic of persistence (A) of a given chemical would indicate that it accumulates in groundwater at a given rate. If we determine the stock-related value of uncontaminated groundwater by conducting a study similar to that reported in Chapter 6, then we are able to assess a chemical-specific tax rate to each unit of sales of the product that will internalise the cost of the implicit use of the groundwater resource. This 'accumulation tax' would then be a combined quantity/quality instrument that would bring the impact of persistence within the decision-making framework of the producer. The producer would then have the incentive to select the aggregate amount of persistence it produces (qA) in a manner that maximises the joint value of chemical production and groundwater quality. A tax only on quantity (and not on expected accumulation) or only on persistence (and not on quantity) will not suffice. In order to regulate the quality of the environmental medium the correct instrument is one that focuses on the product of quantity and quality.

Integrated accumulation taxes

We have shown that, in the case of chemical manufacture, the tax needs to take into account the ability of the manufacturer to substitute between the quantity of chemical sold and its quality of persistence. We shall now investigate how these general principles are affected by the manufacturer's capacity to construct chemicals that react with various environmental media (water, air, living matter). That is, the preceding analysis assumed that the chemical would be designed (by reason of market considerations) with a single type of activity in place (reactivity with fats, water or air); however, it might be the case that the producer could substitute between these media in certain circumstances. For example, a producer might be able to use either the water or the air as the vector for transporting a chemical toward its objective; if only one of these environmental media were taxed, then the incentives to utilise the other would be enhanced. In addition, there also will be distinct levels of harm indicated by use of different environmental media, and the policy instrument must be developed in order to take this variety into account. It is not just the persistence of chemical substances that determines their societal externalities, but also the medium within which they accumulate; hence, accumulation taxes must be developed to anticipate these alternative sources of substitutability.

The choice of environmental affinity

There are three possible environmental media within which chemicals might accumulate: water (e.g. atrazine), organic matter (e.g. DDT), and the atmosphere (e.g. chlorofluorocarbons, CFCs). An integrated approach to regulation must take into account the potential for substitution between these three media. That is, the societal object must endogenise the producer's choice of the characteristic of 'affinity' as well as the choice of persistence.

The effect of this alteration to our framework is to require that all potentially affected resource stocks must be taken into consideration when the optimal regulation is designed. In terms of the simple model laid out above, there are now three resource constraints to be taken into consideration within the societal objective (rather than the single dynamic constraint set forth in (9.3):

Societal objective function (integrated)

$$\max_{\{q,A,\alpha_i\}} \int_0^\infty e^{-rt} [D(W, F, V) + p(A)q - c(A)q]\, dt \qquad (9.7)$$

$$\text{subject: } \frac{dW}{dt} = -\alpha_W A_W q + R_W W$$

$$\frac{dF}{dt} = -\alpha_F A_F q + R_F F$$

$$\frac{dV}{dt} = -\alpha_V A_V q + R_V V$$

where the constraints relate to water (W), organic (food chain) (F) and atmospheric (V) resources respectively.

The integration of accumulation policy

From the above discussion, the accumulation of a chemical in any medium can be separated into two parts: affinity (solubility, relative rates of absorption, etc.) and persistence (rate of degradation). In order to cause producers to make optimal choices regarding the characteristic of affinity, it is necessary for the instrument to be redefined in order to take the choice of environmental compartment into consideration.

The tax per unit of chemical produced must now take into account the accumulation coefficients chosen by manufacturers for each of the environmental media $(A_{W, F, V})$ weighted by the relative affinity of the chemical for each of the various compartments $(\alpha_{W, F, V})$. The tax must also incorporate the social stock values applicable to each of the distinct media and denoted $\mu_{W, F, V}$. The integrated accumulation tax on each unit of chemical production is then:

The integrated accumulation tax

$$I_2 = \alpha_W(A_W\mu_W) + \alpha_F(A_F\mu_F) + \alpha_V(A_V\mu_V) \qquad (9.8)$$

where α_i is the relative affinity of the chemical for the various environmental media (water, organic and atmospheric); A_i is the accumulation rate of the chemical within each of the various environmental media; and μ_i is the stock-related value of one unit of each of the various environmental media (which equals in the steady state the present value of the flow of benefits in perpetuity from an additional unit of stock of the particular resource D_i' discounted at that environmental resource's own rate of discount $(r - R_i')$).

To summarise, optimal policy regarding persistence requires the regulation of the ability of the chemical producer to select both the characteristic of persistence and the characteristic of affinity. This requires that the accumulation tax concept is integrated over all three of the environmental media in order to negate the capacity for substitution between them. Taxing only the quantity of chemical sold will cause the manufacturer to increase the accumulation rate (or decrease the degradation characteristics) of chemicals, resulting in little or no change in the concentration of chemicals in any single accumulation sink. Likewise, controlling the accumulation of chemicals in groundwater alone may do no more than effect a shift of chemicals from one environmental compartment to another. Instead, an *integrated* approach to regulation is suggested, controlling accumulation in all (three) resources. Aggregate chemical accumulation can be addressed only by virtue of controls over: (1) affinity; (2) persistence; and (3) quantities. The integrated accumulation tax (I_2) addresses all three choices simultaneously.

Local characteristics and accumulation taxes

The next complication to be introduced into the analysis is the problem of local heterogeneity. Up to this point it has been assumed that there is a single, undivided stock of the relevant resource over the region to which the policy is to apply. This is not usually the case. For example, in the case of groundwater, acquifers may be extremely localised and completely detached. Then the issue becomes how particularistic the policy-making process should become.

One approach to incorporating important local differences into optimal policy-making would simply be to make the policy jurisdiction coincide with the resource boundaries. This would imply, for instance, the development of as many different accumulation taxes as there are acquifers. The cost of such a particularistic approach is generated by the availability of economies of scale in either chemical production or policy enforcement.

As an illustration of this general point, consider the following scenario. If the design of chemical products and the enforcement of policies was costless, then each community could regulate manufacturers according to the individual characteristics of that community. They would trade off their own anticipated costs and benefits from chemical usage, and determine a local chemical tax accordingly. For example, an agricultural community with an adequate alternative water supply might place a high value on persistence whereas the same community relying upon groundwater might place a much lower value on persistence. Chemical companies would then design chemicals (selecting affinity and persistence) taking into consideration each community's own set of priorities. Farmers would also apply differential rates of chemicals given the differential tax schedule and the differential prices that would result. Each locality would be able to achieve its own first best outcome regarding the trade-off between chemical usage and groundwater contamination.

Of course this scenario can not occur. The design of chemical products is a very costly endeavour, and chemicals cannot be designed and produced in the number of varieties required by the range of characteristics existing in different localities. More realistically, chemicals are probably designed to apply to the average set of circumstances prevailing across a large numbers of markets. Also, the application of differential taxes is problematic with respect to differing localities on account of the mobility of farmers and their capacity to avoid these taxes by means of cross-boundary purchases, especially as the taxes become more particularistic and the boundaries become nearer in proximity (see Andreasson-Gren, 1989, for a related analysis applicable to nitrate regulation). Therefore, it cannot be assumed that user-based responses to particularistic regulation will not also be problematic; users are as likely to avoid a particularistic tax schedule as to adapt to it unless large enforcement costs are undertaken at the local level.

All of these considerations indicate that there is a cost inherent in the use of a more particularistic regulatory regime, and that this cost must be taken into consideration in its construction. There is no avoiding the fact that locally prevailing conditions must be taken into account when construct-ing the optimal policy, but the optimal policy response to heterogeneity might be the introduction of additional types of instruments rather than just more of the same.

This point can be demonstrated within the context of the following modifications to our simple model. Returning to the basic form of the model in equations (9.1) to (9.4), let us now consider the possibility that there are a large number of communities each defined by its reliance upon its own wholly segregated resource (S_i). Each community will then purchase and apply within its district an amount of the chemical (q_i), thereby acquiring a distinctive level of community benefit (V_i) per unit of use (i.e. the broader community may benefit from the use of the chemical in terms of enhanced production, employment and general economic activity). At the same time the chemical accumulates in the local water supply at the rate $(A_i q_i)$ on account of the character of the local substrata, resulting in a loss of commu-nity benefit of the amount $D_i'(S_i)$. Given the particular characteristics of the community (i.e. V_i, D_i, A_i). the first best solution would be for each commu-nity to have its own effectively implemented accumulation tax, resulting in chemical designs (A_i^*) and rates of application (q_i^*) specifically suited to it.

In this model we will assume that this is not cost effective, because of economies of scale in chemical production and policy enforcement. Without loss of generality, I will instead assume that the chemical manufacturer designs only a single chemical for use across all of the various groundwater-based communities. This chemical will be based on the mean characteristics prevailing across these communities $(E(V), E(D), E(A))$, implying a simple accumulation tax of I_1 based on these mean values (as well as the mean stock value they imply). These values in turn imply that the chemical will be designed to accumulate at the producer-selected rate of A (which rate will apply in reality solely to that community with the average geological/hydro-logical conditions). This single type of chemical will then be available across these heterogeneous communities at the price of $(p + I_1)$ still assuming that the chemical industry is competitive in nature (and hence passes on the full amount of the accumulation tax to the user of the chemical).

Societal objective function (heterogeneous conditions)

$$\max_{\{q_i\}} \int_0^\infty e^{-rt} [D_i(S_i) + (V_i - p - I_1)q_i] \, dt \qquad (9.9)$$

$$\text{subject:} \frac{\partial S_i}{\partial t} = -A_i q_i + R_i(S_i)$$

$$V_i - p - I_1 \geq 0$$

In this construction, the society concerned is the local community attempting to regulate the use of chemicals in the context of its own special circumstances. There has previously been applied a producer-level tax in order to cause the producer to consider average circumstances in its design of its product, but there remains the necessity of considering how each local community will adapt to the combination of product characteristics and prices that will result. What sorts of user-level instruments are required in order to address the residual problem of heterogeneity? The conditions for the optimisation of the objective set forth in expression (9.9) indicate the nature of the solution.

Static maximisation conditions (complementary slackness)

either:
$$V_i - A_i \mu_i - (p + I_1) < 0$$
$$q_i = 0$$
$$\qquad (9.10)$$
or:
$$V_i - A_i \mu_i - (p + I_1) = 0$$
$$q_i > 0$$

These conditions imply the use of an additional instrument at the local level. There are three possibilities suggested by the conditions within expressions (9.10).

Local level regulatory instruments (user taxes and zoning)

(Case 1) $V_i - (p + I_1) < 0$: Here the producer-level tax negates any incentives for local chemical use. In this case the local community will not purchase the chemical simply because the producer-level tax has rendered it economically unsuited to the area.

(Case 2) $V_i - (p + I_1) - A_i \mu_i > 0$ (over sufficient range of q_i): Here there is an adequate benefit from local chemical usage to justify its introduction and further regulation. In this case a supplemental tax of $I_3 = (A_i \mu_i - I_1)$ is charged on each unit of the chemical applied within the locality.

(Case 3) $V_i - (p + I_1) > 0$ *and* $V_i - (p + I_1) - A_i \mu_i < 0$: In this intermediate range there are incentives under the producer-level tax to introduce the chemical within the locality, but inadequate additional benefits to justify another level of complex regulation. In this case quantity-based restrictions ('zoning') may be introduced in order to avoid both chemical accumulation and non-economical regulation. The introduction of this instrument is based on the assumption that it will be less costly to distinguish between different types of chemical in users' hands than it would be to distinguish between the point of acquisition of the same chemical.

From this analysis we are able to see that the optimal regulatory system will be designed in order to address the incentives of both producers and users. Producers must be regulated to take their choices concerning chemical design in a societally optimal fashion. These choices will very likely be taken only once for a given chemical product. Chemical users must then be regulated in order to take the special circumstances of local conditions into account in the use of these mean-calibrated chemicals. These use restrictions may take the form of either supplemental taxes or (where the gains from use cannot justify the costs of the additional regulation) local prohibition ('zoning').

Further complications: dynamic and strategic considerations

There are two further complications that must be introduced into the analysis. One concerns the period of time prior to reaching the steady state that has been discussed thus far. The other concerns the problems arising from the simplistic industrial structure that has been modelled.

Regulating for strategic industrial responses

So far the model that has been discussed has been kept as simple as possible in terms of its industrial structure; the assumption throughout has been that the chemical industry operates in a competitive manner. This assumption is of fundamental importance in determining the reaction of the chemical producers to regulation. In our analysis the producers respond non-strategically (in the desired manner), allowing for the discussion and description of simple instruments to proceed.

The problem with this depiction is that it is not grounded in reality. The chemical industry is one of the most concentrated industries that exists at both national and global levels. Not only is the industry concentrated, but (on account of the availability of chemical patents) much of production takes place in a monopolistic environment rather than a competitive one. Firms in such an environment will respond strategically rather than automatically to regulatory policies.

Analysis of this issue (see Mason, Chapter 8, this volume) indicates that the form of the strategic response is predictable. In response to governmentally introduced resource scarcity, chemical firms (in either oligopoly or monopoly settings) will attempt to hoard the resource in order to prevent potential and existing competitors from having access to a necessary input. That is to say, when governments attempt to institutionalise the recognition of groundwater as an input into the chemical industry, and price it in relation to its scarcity value, then oligopolistic chemical firms will have the perverse incentive to *expand* consumption of the groundwater resource in order to prevent other firms from making access to that resource. Analogously, if governments introduce aggregate limits on specific chemical products (as in the case of product-specific bans), then firms will also have the incentive to exhaust the resource with regard to this product rather than conserve it.

In short, the government introduction of resource scarcity where none has existed before will create incentives for both conservation and appropriation within the context of an oligopoly. This implies the need for a modification to the basic accumulation tax. A premium must be added to the accumulation tax in order to remove the incentives for hoarding the resource.

Regulating the transition phase

One further remark is necessary to place the preceding analysis into its appropriate context. The analysis within this chapter has been dynamic in nature but focused upon the 'steady-state': the point in time at which the desired water quality has been attained. All policy after that point is directed towards the maintenance of a given standard, and this implies the restriction of the flow of chemicals into groundwater to a given amount per period. This allows the creation of a single time-invariant accumulation tax for this purpose.

Until that point there are two distinct policy paths to the optimal steady state policy. One possibility is simply to leave chemical use unregulated until the time where regulation is required in order to maintain the steady state standard (equivalent to the most rapid approach path). The alternative is to commence with a low value of accumulation tax, gradually increasing it to the steady state value over time (the tax rate in this instance should be raised at a rate equivalent to the resource's 'own rate of discount' $(r - R)$). This path to the steady state is defined by the portfolio balancing condition presented in equation (9.5). The choice between these paths is meant to depend upon the existence of non-zero elasticities regarding the primary control variables (q, A) within the societal objective function. If such non-linearities exist, then there is good reason to provide a smooth transition phase into the steady-state policy; if not, then the optimal approach is simply to move directly to the steady-state policy ('bang-bang equilibrium').

In general non-linearities will exist with respect to the control variables and this implies the importance of a 'phasing-in' period. The optimal accumulation tax will in fact be a tax schedule rising over time towards the optimal tax that will be applied in the steady state (when groundwater quality is maintained constant).

Conclusion: application of optimal policies within the European Union

The optimal policy derived within this analysis finds that it would be best to apply a global accumulation tax schedule designed in regard to the mean set of characteristics prevailing within the regulated community, with supplementary taxes (and quantity restrictions/zoning) to be assessed by individual localities where accumulation and/or harmful effects exceed the mean.

This sort of regulatory approach makes sense within the existing legal framework of the Union. EU law has allowed stricter environmental standards to be applied under certain circumstances but not more lax ones. The reason is obvious: for competition to be played on a 'level playing field', certain 'mean-related' policies must be introduced in order to prevent policy-induced non-competitiveness. Hence, European Court rulings have allowed stricter environmental enforcement, but not more lax environmental enforcement than the EU standard. Supplementary taxes allow individual states to factor some local conditions into the policy choice, without inflicting any additional trade disadvantage upon fellow EU Members. This is consistent with the Court rulings regarding the allowance of 'stricter derogations' from EU environmental standards. These are allowed so long as they are not used to restrict trade between Member States; in this case, the handicap would only be self-inflicted and hence not in violation of any uniformity requirements.

Therefore, since EU law requires that environmental standards be standardised, it makes sense to use the mean characteristics within the community to determine the Union-wide tax schedule. However, since the optimal policy requires some heterogeneity, this will require that the tax schedule be supplemented by locally implemented and enforced tax premiums. In this fashion, approximately the optimal amount of standardisation and heterogeneity may be introduced.

Appendix: Measures of chemical fate

Decay rates

Half-lifes (photo, hydro, bio): This is the length of time during which the mass of the substance is reduced by one half in the various environmental compartments. Decay occurs through reactions with sunlight (photo), water (hydro) and/or bacteria (bio).
Mineralisation half-life states the amount of time required for degradation to carbon dioxide and water.

Affinity quotients:

There are two different sorts: quotients and saturation thresholds.

Quotients: relative affinity of substance for organic compounds and water:
K_{ow}, relative affinity of substance between simulated fat (n-octanol) and water (i.e. if offered the opportunity to react simultaneously with either, the proportion of substance that will bond with either substance).

K_{oc}, relative affinity of substance between water and organic carbon in soil (same idea as K_{ow}):
Also, some substances will also bond with the inorganic components within soil (this gives a ratio K_p that represents propensity to bond with all parts of soil).
Saturation thresholds: the capacity of environmental media (air and water) to absorb the chemical.:
Water solubility: the amount of the substance that can be dissolved in one litre of water.
Vapour pressure: the amount of the substance that can be absorbed by air (measured in terms of vapour pressure in steady state).

References

Andreasson-Gren, I.-M. (1989). *Costs of Controls on Farmers' Use of Nitrogen*. Economic Research Institute, Stockholm.
Baumol, W. and Oates, W. (1979). *The Theory of Environmental Policy*. Cambridge University Press, Cambridge.
Conrad, J. and Olson, L. (1992). The economics of a stock pollutant. *Environmental and Resource Economics*, **3**, 245–58.
Cropper, M. (1976). Regulating activities with castrastrophic environmental effects. *Journal of Environmental Economics and Management*, **3**, 1–15.

10 An analysis of alternative legal instruments for the regulation of pesticides[1]

Michael G. Faure and Jürgen G. J. Lefevere

Introduction

This chapter is a follow up to a previous paper[2] in which a first attempt was made to address some of the institutional aspects of the European Commission Directive on drinking water[3]. Whereas the previous paper, after a short theoretical introduction, focused mainly on the legal history of the Directive and its implementation in Italy, this chapter is devoted to the more theoretical question of what legal instruments can be used to avoid an accumulation of pesticides in drinking water. The current situation with drinking water is that the EU is using Directives to set strict quality standards to protect drinking-water supplies. Once these EU standards are set, the Member States have to implement them in their national systems. It has been shown that the current regulatory approach for pesticides is not functioning satisfactorily. The research by the ecotoxicology group has demonstrated that the current Italian approach of a ban on one pesticide (i.e. atrazine) is inefficient since it does not prevent an accumulation in drinking water of other pesticides. Moreover, the research has shown that the real problem with pesticides is not their toxicity, but more importantly the accumulation of potentially toxic elements in drinking water[4]. The current regulatory approach gives the wrong incentives to pesticide manufacturers and to the users of the pesticides, the farmers. It does not lead to a reduction of toxic pesticide accumulation in drinking water.

The main research question that has therefore to be addressed in this chapter is 'what optimal legal rules should be developed in a combined European and national legal system for preventing the accumulation of pesticides and their toxic ingredients in drinking water?'

In order to answer this question we will first have to answer some other specific questions. First, we will have to establish the precise goal that we want to reach with the rules. Once we have established that goal we will try to establish which incentives need to be given to whom; in other words, who will be subject to obligations under the regulation. Once the addressees of the regulation have been determined, the optimal differentiation of the regulation has to be examined. The question is whether a general rule suffices or whether the rule has to be differentiated according to possible location-specific circumstances that play an important role in determining the regulation. Once this level of optimal differentiation has been established the instrument of regulation has to be chosen. We will examine different instruments that can be used to regulate the use of pesticides. The conclusions of this analysis will be used to examine how these rules should be set in a combined European and national legal approach, taking into account an optimal distribution of powers, so that the subject can be regulated most effectively.

The chapter is constructed as follows. After this introduction some disadvantages of the current regulatory approach are outlined. Then we try to identify the precise aims we want to achieve with the regulation. After this we turn to some general questions on how standards should be set. These findings are then applied to some legal instruments. Subsequently, we describe the specific properties and advantages of setting rules in an EC context. The final question that is addressed is how these optimally differentiated rules should be set within a combined European and national context. Finally, a few conclusions are drawn.

Some problems with the current regulatory approach

The European Union Standard

Early on in the research project, all participants in the group came to the conclusion that the current regulatory approach with respect to pesticides is inefficient. This could apply to the 0.1 µg/l standard for pesticides in the drinking-water Directive, but especially to the Italian ban on atrazine. The problem with the current EC standard of 0.1 µg/l is that it is a rather restrictive interpretation of toxicological knowledge. In constrast to the EU norm, the World Health Organization uses the considerably higher limit of 2 µg/l. It has been argued that the EU standard is too strict for various reasons. Vighi and Zanin for instance argued that in terms of real risk for the exposed human population as a consequence of the consumption of water contaminated by atrazine, it is clear that the EEC limit for individual pesticides in drinking water is very conservative and, in many cases, orders a magnitude [sic] below a demonstrable threshold value. The philosophy of this part of the EEC Directive is based on the notion that products such as pesticides (which are biocides by definition) should not be present in drinking water because this is a fundamental resource, intended for regular daily consumption.[5]

The EU standard, which was basically set at the lowest amount detectable at that time[6], now causes serious problems since the methods of analytical detection have improved, so that in many cases the 0.1 μg/l norm is violated although not causing a serious threat to human health. One could, however, argue that the 'zero tolerance' attitude of the EU takes a broader approach to the problem of drinking-water pollution than does the WHO. The philosophy of the EU is based mainly on broader principles. First, it is argued, since pesticides are toxic substances by definition, they should not be present in the natural environment at all. Secondly, the EU approach is based on the view that drinking water is an all-important human resource which human beings use intensively for a lifelong period. The possibility of high exposures therefore in the view of the EC warrants a strict approach.[7,8]

The EU standard causes serious problems both for states and for water companies. States can be held liable by individuals for not, or incorrectly, transposing the European Directive into national law.[9] They risk a conviction by the European Court of Justice if the water quality in their country does not meet the requirements of the Directive[10] and they can be forced to pay a penalty payment for as long as the conditions of the Directive are not met.[11,12] Since the Directive must also be implemented in national legislation, the water companies supplying water that does not meet the requirements of the Directive risk civil liability and their officials even risk criminal liability on the basis of national law. The effect will be that both the state and the water companies will have strong incentives to invest in treatment methods in order to meet the 0.1 μg/l EU limit and so avoid liability.[13] Although one could argue that more resources are being spent to reach the EU standard than would be efficient, one could defend the EU approach even if the attainment of a stricter standard would be more costly. Relevant in this context is, however, also how much Member States are willing to pay for clean drinking water, which is of course dependent on their varying preferences.

The Italian approach

At first sight the Italian approach to the pesticide problem, i.e. the ban on one compound (atrazine), looks effective and cheap to enforce. In practice, however, it clearly raises several problems. It is a very harsh, uniform rule that does not take local differences into account in any way. A ban on atrazine might be useful in one region of the country, where products are grown that need a lot of pesticides and where the hydrogeological situation will lead to a high accumulation of pesticides. The simulation model constructed by Zanin *et al.*[14] shows that, in areas with permeable soils, under rainy weather conditions even small amounts of atrazine have a significant effect on the levels reached in drinking water. In other, drier regions, with less permeable soils, a ban on atrazine might not be necessary at all. In those regions the agronomic benefits of the use of atrazine might largely outweigh possible disadvantages for the pollution of the drinking water. By introducing such a general ban on atrazine, including in those regions where its use would be harmless to the water consumer and beneficial for the farmer, Italian farmers are put in a disadvantageous competitive position compared to farmers in countries where a (restricted) use of atrazine is still allowed. Furthermore restrictions on the import of atrazine might cause market distortions within the EU.[15]

Generally, the ban on one compound is ineffective if it can easily be replaced by substitutes that also accumulate in drinking water. Farmers will shift to other pesticides for the protection of their crops,[16] which will then need new legislation to ban or regulate their use. If this is not done, then other pesticides will be found in amounts exceeding the EU limit, which is indeed the case in Italy today. Hence, the total ban on atrazine does not solve Italy's regulatory problem of non-compliance with the EU limit. As Swanson has argued, the ban does not provide the pesticide manufacturers with incentives to produce pesticides that do not accumulate in drinking water.[17]

A basic problem with the current regulation of the use of pesticides in this context is the fact that it is not the pesticide toxicity as such that is the problem, but its ability to accumulate in different compartments of the natural environment, as in the case of atrazine in drinking water. The accumulation of atrazine in drinking water depends not only on various chemical characteristics (among which the persistence of the pesticide is important) but also on the types of product grown (which influences the demand for pesticides), on the amount applied and on hydrogeological factors. The conclusion would be, therefore, at first sight that the legal system should take all these factors into account. The question we have to address now, therefore, is whether a regulation should be uniform, or whether there can be differentiation, taking these factors into account.

Regulating pesticides: goals and incentives

Before discussing the general principles of standard-setting and giving some examples of regulatory instruments, the goal of pesticide regulation has to be identified; in other words, 'what is the problem that we want to solve?' The current problem is the presence of potentially dangerous substances, *in casu* atrazine, in the drinking water of the Member States. Although there is some discussion on the admissible level of atrazine in drinking water,[18] it is generally agreed that the amount of atrazine in the drinking water should not exceed a certain limit. Therefore a general quality standard for the pollution of drinking water by pesticides should be established, either as a legal norm or as a policy objective.

As we have seen, our problem with pesticides is their accumulation in drinking water. In order to overcome the accumulation problem, the causes of this accumulation have to be found. There is no general rate of accumulation for a unit of pesticide. The accumulation rate of atrazine is dependent on several factors. Atrazine is water-soluble and therefore readily accumulates in drinking water. Because of this water solubility, weather and hydrological conditions play an important role. The accumulation of atrazine in drinking water is much higher when the pesticide is applied in wet conditions than when it is used in dry conditions. The geographical characteristics of the region of application also play an important role. Areas with permeable soils are more likely to become contaminated than areas with, for instance, clay soils.[19] Thus, widespread drinking-water contamination by atrazine in the Po Valley in northern Italy has been caused by the infiltration of water with pesticides through very permeable soils[20] into the underlying aquifer.

As the accumulation of atrazine is very dependent on local geographical and hydrological conditions, two factors can be identified that determine the actual pollution of the drinking water by atrazine. First, there is the actual accumulation rate of the pesticide combined with its toxic properties. Secondly, there are the geographical and hydrological conditions under which atrazine is being used. The aim of regulating the use of pesticides in general and of atrazine in particular should therefore also be two-fold. First, incentives should be given to producers of pesticides to produce 'less polluting' alternatives for atrazine. Secondly, because research has shown that there are at present no good alternatives available to replace atrazine,[21] the aim of regulation should be to give farmers incentives to use atrazine under the right conditions in the right quantities on the right soil.[22] In the following sections the methods which can be used to reach these aims are examined.

Principles of standard-setting

In the previous section we outlined some aspects of the goal of the legislation and the incentives that this legislation would have to give to the farmers and the pesticide producers. The question now arises as to how these criteria, especially location-specific features such as the hydrogeological situation, can be taken into account in the process of regulating the use of pesticides. In the next section we address the question of what kind of legal instrument would be best to take these criteria into account. In this section, however, we first formulate a few general observations concerning standard-setting in environmental law that might be also of relevance for the pesticide problem.

Setting the level of environmental protection

A crucial question is how, in a public interest perspective, efficient environmental standards should be set. It is often argued that the standard should be set at the level where the marginal costs of pollution abatement are equal to the marginal benefits in reduction of environmental damage. This means that a newly developed environmental technology would only be efficient and should therefore only be incorporated into a legal standard if its marginal costs are lower or equal to the marginal benefits in additional reduction of environmental damage[23]. In principle this can also be applied to the use of pesticides. The optimal level of care or the appropriate efficient standard would then address the type of pesticide to be applied or the amount or mix of various pesticides under specific hydrogeological conditions. To determine the optimal standard for the pesticides to be applied, one would have to take into account the benefits of the use of agriconomically sophisticated pesticides with a high kill rate of specific weeds at low costs on the one hand versus the marginal costs of the additional use of the pesticide for the environment on the other. An estimation of these costs, including the costs of subsequent additional water treatment, has been made by the economist group.[24]

Differentiation and optimal specificity

Obviously these marginal costs of the additional use of pesticides will depend upon the extent to which the pesticides accumulate in drinking water. As we have just indicated, the property of accumulation is dependent not only upon the type of compound used and the amount applied, but also upon local and hydrogeological conditions. The marginal costs of the use of some pesticides might therefore be relatively low in some regions where the accumulation risk is low as a consequence of the specific hydrogeological condition.

Hence, the question arises as to whether a standard for the use of pesticides should be uniform or differentiated. Generally, economists argue that, in a public interest view, standards should be differentiated not only according to region, local needs and industry type, but also according to preferences of the public. In different regions citizens might have different preferences regarding the appropriate trade-off between environmental quality and industrial production.[25] The question of course arises as to what influences the choice between uniform or differentiated standards and, if differentiated standards are preferred, through what kind of legal mechanism should they be introduced and enforced.

The question of optimal specificity has received some attention in the literature on law and economics, especially in an article by Ehrlich and Posner, but also recently in the work of Ogus.[26] Uniform standards are obviously cheaper to formulate and enforce. They also do not depend on individual bargaining between an administrative agency and those regulated, so that there is less risk of capturing[27] of the agency[28]. On the other hand, it is clear that the social costs arising from an activity will differ according to location specific circumstances. Uniform standards that do not take into account these differences will therefore lead to welfare losses. Moreover, individuals in various regions might have different preferences with respect to environmental protection.

These are arguments that take into account location-specific circumstances, where possible. It is, however, also clear that one could go very far in differentiation by taking all kinds of details into account, which would lead to highly varied environmental standards. Although the optimum in an ideal world would be to set the environmental standard exactly equal to the location-specific costs, it goes without saying that this is impossible because of the administrative costs that are incurred with a highly detailed standard-setting process and because of the information and enforcement costs that might also increase with the greater differentiation of standards. Hence, there is a trade-off between the benefits of adapting the standard to the specific social costs of an activity by introducing differentiated standards on the one hand and the increasing information, administrative and enforcement costs of such an increased differentiation on the other. The trade-off should result in an optimal specificity of standards, found at the point where the additional administrative costs of a further differentiation equal the benefits of such a differentiation. In the words of Posner, 'the question is whether the benefits of particularization outweigh the costs'.[29]

The costs of differentiation

Let us address briefly the various costs involved in a further differentiation with respect to the pesticide accumulation problem. We mentioned above that the real social costs of the use of pesticides will vary according to their propensity to accumulate in drinking water. This accumulation problem again depends upon several factors, among which are the types of pesticides used and the amounts applied to the soil, the types of product grown on the land and the hydrogeological conditions.

What are the costs involved in taking into account some of these differences? Three types of cost can be distinguished here. Of primary importance is the *information* cost[30]. This is the cost incurred in collecting the information necessary for the application and elaboration of the rule by either the standard setter or the regulated. A second type of cost is that due to *administration*. Once a standard has been set, a rule has been made, and this rule has to be applied. In the case of licensing, for instance, authorities will have to set up a body that gives the licenses. Apart from applying the law, the law also has to be enforced. This enforcement produces the third type of cost, *enforcement*. The authorities have to set up some kind of environmental police force to check compliance with the legislation by the people to whom this legislation is addressed.

The *information* cost to examine what types of pesticide are highly persistent and will certainly accumulate in drinking water and the accumulation effects in cases where mixtures of pesticides are used is at first sight enormous. For an individual farmer it would be extremely costly to acquire that kind of information. Of course, some of this information on the persistence or accumulative properties of pesticides is readily available through technical journals. If not, it is clear that information on these kinds of property can be better acquired by a government, after which the information can be passed on to the public at large. The advantage of a governmental examination is also that economies of scale advantages apply with respect to research and development costs. Once the specific properties of a pesticide have been found out, the information can be made available and the cost can be spread over a larger number of people. The same goes probably for investigations into the hydrogeological conditions of the soil. The effect that the condition of the soil has for the accumulation of pesticides in drinking water can scarcely be examined by one individual farmer. A further problem would be that if an individual farmer invested in research to examine the hydrogeological conditions of the soil and then made this information freely available, third parties could take advantage of his or her investments without making any compensating payments. The only type of information that is probably readily available is what types of product require a large amount of a certain pesticide to produce an economic yield. Farmers tend to be fully aware of what types of product do not need so much in the way of persistent weed killers or other pesticides.

If one or more of these decisive factors were to be incorporated into legislation, information costs will also be incurred with respect to administrating and enforcing a certain regulation. In the next section we examine several legal instruments and ask the question 'what legal instrument would be best adapted to take into account those specific conditions that influence the accumulation problem'. An important factor in that respect is of course whether the information needed to apply a certain legal instrument can be obtained by the regulatory authority at relatively low cost. Hence, the information needed will also influence the choice of legal instruments to reach the required differentiation.

The *administrative* cost varies, of course, according to the legal instrument chosen. We discuss this point further in the next section. The same can be said for the *enforcement* cost.

Interest group aspects

Within this socio-legal framework of standard-setting and regulating the accumulation problem with respect to pesticides, one should note that the choice of legal instruments will not only be influenced by the various costs mentioned, but also by the politico-economic dimensions. Standards can of course be influenced by private interest groups. An obvious disadvantage of an overdetailed differentiation is that the potential for introducing specific private interests will increase.

An example of detailed European legislation that appears to have been influenced by private interest groups is the Directive of 22 March 1982 concerning discharges by the chlor-alkali industry[31]. It is indeed remarkable that a separate regulation was promulgated for this specific industry. The text preceding the Directive states that the pollution by discharges of mercury into water is caused, to a large extent, by the electrolysis of alkali chlorides. It says: 'in the first instance limit values should be established for this industry and quality objectives should be laid down for the aquatic environment into which mercury is discharged by this industry'. Thus, it would undoubtedly be useful to set quality objectives for the aquatic environment and it might be necessary to set limit values as well. As far as limit values are concerned, however, the question arises as to why the EU should set these limits and in particular one can wonder why separate standards are set for one sector of industry. Another example of the increased private interest influence of detailed legislation is the titanium dioxide Directive. This Directive incorporates, in its Articles 4 and 6, a classic 'grandfather' clause[32]. The Directive sets different standards for firms that are already in the market on the date the directive is promulgated than for newcomers to the market. Obviously the established firms can continue the use of their existing treatment plant whereas newcomers have to comply with more severe standards. One cannot therefore escape the impression that these types of specific standard-setting aim more at rent-seeking (i.e. obtaining an economic advantage) through limiting market entry than at genuine environmental concerns. If the latter were the case a separate Directive for the chlor-alkali industry would not be necessary at all in the first place.

The finding that some Directives create different standards for small branches of industry such as the chlor-alkali or the titanium dioxide industry cannot be explained as a useful means of differentiation in the public interest. It seems better to fit into the public choice theory that predicts that especially small, well-organised pressure groups with low organisation costs, no start-up costs and a single issue to fight for, will be successful rent seekers.[33] Hence, within the choice of legal instruments one should also take into account that, preferably, a legal framework should be so designed as to minimise the risk of successful rent-seeking by lobby groups.

In the next section we apply these general observations to the choice of legal instruments by asking the question 'what type of instrument would be best suited to prevent pesticides from accumulating in drinking water?'

Choice of instruments for environmental protection

Generally three categories of instruments for environmental protection can be distinguished. The first category consists of *liability rules*. By setting general rules for liability for environmental pollution, the legislator can steer control of the process of environmental pollution by, for instance, placing strict liability on the polluter, or shifting the burden of proof. The question of liability in a specific case will often be made by the judge. The second category of instruments are the *market based* or *economic instruments*. These instruments are characterised by minimal intervention by the legislator. The legislator merely sets up a framework in which the 'invisible hand' of the market economy will eventually lead to the right result. The third type of instrument is the *command and control* approach. Using these instruments the legislator intervenes directly in the polluting processes by, for example, requiring firms to have emission permits or product licenses. These three categories will be examined below.

Liability rules

In our previous paper on the institutional aspects of the European Commission drinking-water Directive we discussed the classic literature on the criteria for safety regulation, as developed by Shavell[34]. In this litera-ture, the criteria for regulation are discussed, especially in comparison with liability rules. Also the information, administrative and enforcement costs are introduced as a criterion. In that respect it is argued that liability rules would be preferred if the necessary information were better obtained by private parties in an accident setting than by a regulatory agency. It seems very difficult to use the liability system to deter an accumulation of pesticides in drinking water. As Swanson indicated in his chapter on the accumulation problem, a liability rule will probably not have the desired deterrent effect since defendants might be judgement proof, or a proof of negligence or of causation might be very difficult.[35] Most literature argues that, with respect to environmental standards, liability rules alone will not have enough deterrent effect to reach a full internalisation of the environmental harm. Moreover, empirical research by Dewees has showed that in the USA the quality of the environment has been improved mainly as a result of regulatory efforts, and to a lesser extent by tort law actions.[36]

To return to the information problem mentioned on pp. 253–4, it does indeed seem inefficient to hold a farmer liable for pollution of drinking water if he or she lacks information on the properties of pesticides applied or on the hydrogeological properties of the land. That information can better be obtained through regulatory agencies and be passed on to the regulated farmer in regulation prescribing or prohibiting the use of certain pesticides in specific regions. Moreover, one should not forget that the information needed to make an efficient use of the tort system is not only the information needed by the potential injurer (i.e. the farmer) which should lead him or her to rational decisions concerning pesticides use, but also information needed by the courts. In a perfect negligence liability system, a court would balance, on the one hand, the various pesticides the farmer could have used and, on the other, the social costs that each one of them might have caused to decide upon the negligence of the farmer. The information needed for a court to engage in such a detailed cost–benefit balancing is enormous.

However, most authors argue that the tort system can still play a role in a regulatory world, for instance to provide additional deterrence in situations where regulation fails.[37] But for the accumulation of pesticides in drinking water it is doubtful whether the tort law system can play even this supplementary role. The information costs for judges seem simply too high to make an adequate cost–benefit analysis to determine whether the farmer used efficient care by applying a certain type of pesticide on a specific product in an area with a given hydrogeological condition.

Economic instruments, the accumulation tax

Swanson mentioned that the aim of pesticide regulation should be to provide the pesticide manufacturers with incentives to produce pesticides that do not accumulate in drinking water. He argues that since 'prevention is better than cure' an *accumulation tax* should be introduced to give manufacturers the correct incentives at the stage when new chemicals are being designed.[38]

The introduction of such an *accumulation tax* could be seen as a first choice solution. Economists have indeed often advanced the imposition of a Pigouvian tax to control externalities. Although this tax approach is supposed to have many advantages over regulatory approaches, it is not yet used to a large extent in many countries. Posner gives some explanations as to why, in general, a pollution tax is so rarely introduced despite being much favoured by economists[39]: a tax might be counterproductive in situations where the victim is the cheapest pollution avoider. This also raises questions about the accumulation tax. Although it might be very costly for the real victims (the users of the water) to reduce pollution by purifying the water,[40] the question certainly arises as to whether the accumulation tax should be levied upon the manufacturer of the pesticide or upon the farmer who applies the pesticides. In order to reach an optimal use of the right pesticides on the right land (with the right hydrogeological conditions) it might be more logical to tax the use of the pesticide, rather than the production. Eventually one could think of taxing both the manufacturer and the user. Furthermore Posner[41] indicates that there are billions of emissions of pollutants every year and it is very difficult to estimate the social cost of each one for the purpose of setting the correct tax rate. Also the social costs of pollutants in different parts of the country are not uniform. Since the correct tax should be equal to the marginal social cost of one pollutant, the information cost required to apply a tax schedule is very high.

These arguments that explain in general why so little is made of taxes on pollution, even though this seems a first choice solution, might also be applicable to the accumulation tax approach.

Command and control regulation

Clearly, therefore, one has to turn to a regulatory solution to take into account the accumulation problem. What are the costs involved? The administrative and enforcement costs are always higher with regulation than under a liability system.[42] It is therefore important to look for a regulatory framework that encompasses the accumulation problem with the lowest possible costs. Four regulatory instruments are discussed here: the prohibition of pesticides, a licensing system, a zoning system and quality standards.

Prohibition

If the propensity to accumulate in drinking water is influenced to a large extent by the type of pesticide used, a simple solution would be to regulate the types of pesticide to be used by farmers *ex ante* by prohibiting the use of certain pesticides that are most likely to accumulate in drinking water. This naïve solution seems, however, inefficient. A property of a successful pesticide is precisely its capacity to persist, so that it needs to be applied less frequently. This inevitably creates an accumulation problem as well. Moreover, by prohibiting as a general rule *ex ante* all pesticides that might accumulate in drinking water, one might also exclude pesticides with good agronomic properties, and thus inefficiencies arise. Again, it might be efficient to exclude the use of these pesticides in one region, but maybe not in another where the hydrogeological situation will not as easily cause accumulation problems. In addition, the Swanson indicated that a simple ban on a certain pesticide will create the wrong incentives for the manufacturers of pesticides.[43] Given the problem of substitution, one would have to ban not only one pesticide but all possible pesticides that could accumulate in drinking water. Given the impracticability and inefficiencies that this might create, a more balanced approach seems warranted.

Zoning

The use of zoning is a regulatory solution that takes into account local conditions influencing the possibilities of accumulation of pesticides. Zoning is a legal instrument used in urbanisation law.[44] It is an instrument that indicates in what part of a city or region specific activities can be undertaken or what types of construction can be built. Classic examples of zoning occur within cities where a specific area is reserved for shops, another one for apartment buildings and similar types of housing, other regions for suburban luxury houses and again other areas for industry. One could of course consider a type of zoning for pesticide use, provided that the region where the pesticides are applied is indeed a decisive criterion for the accumulation problem. If hydrogeological conditions which are crucial for the accumulation problem can be easily recognised by an administrative agency functioning as a zoning board, the agency could indicate in what area certain pesticides could be applied. Such ease of recognition would also solve the information problem. In any event the information costs for an individual farmer to examine the hydrogeological situation of the soil on which he or she is growing products are of course much higher than the costs of an agency that can do the same research for a whole area. The administrative costs of zoning are still relatively low once the relevant geological conditions are known. In that case the zoning board would probably issue a regulation indicating that certain types of pesticide are banned in areas with a sensitive hydrogeological condition. A decree of the zoning board can be made self-executive and enforced by criminal sanctions. Of course it involves enforcement costs to control such a decree, but clearly the same problems arise with the general ban on atrazine in Italy that also has to be enforced.

The difference with the current Italian ban on one compound would be that, with zoning, a more sophisticated analysis could be made of the influence of several compounds, given the hydrogeological situation, and a board of technical experts could decide that the use of several compounds in a specific area is prohibited. This would also have the advantage that the substitution problem is solved, since the exclusion would apply not just to one pesticide but to all pesticides capable of accumulating in drinking water (provided of course that this is not true for all pesticides and that the accumulation properties of the pesticides can be easily recognised). Another advantage is that the efficiency of excluding agronomically useful pesticides is not extended to regions where the use of these pesticides would be relatively harmless. Thus, the costs of an exclusion can be limited through zoning and there should be some efficiency gain compared to the current situation in Italy.

Of course zoning creates politico-economic problems. In particular the discriminatory nature of zoning will mean that farmers in less sensitive hydrogeological areas receive a benefit since they can continue the use of the agronomically sophisticated pesticides. One could, for political reasons, consider some kind of compensation to the 'losing farmers'. However, one could argue that a farmer should always take into account the local conditions such as weather and the properties of the soil when deciding which crop is to be grown. Therefore one could argue that zoning, which takes into account the accumulation problem, should in the end lead to the correct incentives for farmers either to change their production to a type of crop that needs less or other pesticides or to relocate crops or the entire farm to areas that are less sensitive for the accumulation problem. A steady reduction in the annual compensation payments may assist this process.

Of course one could also suggest that the zoning ordinance was not aimed directly at the types of pesticide used, but at a prohibition of the cultivation of a certain product that demands an excessive use of pesticides. However, since the pesticide used is the real problem it seems better to direct a regulatory action such as a zoning ordinance towards such a use. Moreover, from a political point of view it might be very difficult to tell a farmer by regulation what type of product can be grown in a certain area. The same effect can of course be reached in an indirect way by prohibiting certain pesticides in hydrogeologically sensitive areas, which should also give the farmer an incentive either to change the product grown to products needing less pesticide or to relocate.

Of course we assume that there is a zoning board which follows a public interest-type balancing of costs and benefits to set restrictions upon land use. However, in reality zoning boards often appear to be political institutions that tend to introduce restrictions on land use in order to increase the value of the property of the people who elect the board.[45] Thus, zoning will prove to be inefficient if these kinds of classic public choice effect take place.

Quality or target standards

Target standards are often used to translate the policy goal of environmental legislation directly into legally binding norms. As they are simply a translation of the policy goal, the information costs of target standards are generally quite low. The problem of quality standards is, however, that they are very costly to adminstrate and enforce upon polluters. Once the quality of, for instance, a river is below a certain level, it is very difficult to find out who is responsible for this pollution, especially when there are several polluting industries located on this river, each making a different contribution to the harmful effect. Target standards are therefore mostly not used as single regulatory instruments, but in combination with other instruments; that is, specific emission standards (licensing system) or economic instruments, where the total emissions of the granted licenses or pollution rights are subject to the maximum admissible amount laid down in the quality standard. In this context, the quality standard is therefore often generally seen as a standard that is legally binding upon the agency that sets the more specific standards rather than upon the polluters themselves.

Licensing

Finally the regulation could go one step further, in the direction of an individual licensing system. In that case either one would incorporate the relevant criteria for the accumulation problem into an already existing environment or other permit that a farmer needs to have or one would introduce a new licensing system. The advantage of the system is that it allows as much differentiation as one wants. A well-informed agency could set the standard for the pesticide to be applied, taking into account all the relevant location-specific criteria. In principle such a detailed and differentiated system could be efficient. Of course it could also go so far as to regulate in detail, for instance, what kinds of product a farmer could grow, but this seems politically infeasible. The administrative costs of an overdetailed regulation might also be too high. Of course, there is a problem of information costs. Will a local agency have all the necessary information to decide what the optimal mix of pesticides to be applied is, taking into account all the technical criteria such as the hydrogeological situation? This might still be relatively easy to determine for a larger area by, for instance, a board of experts in a zoning system, but can be relatively difficult to determine for local administrative agencies that have to issue the licenses. In that respect one could combine a zoning system with licensing, whereby the licensing administrative agency is bound by the zoning decree. But then of course what are the additional benefits of licensing? Unless one really wants to regulate in so much detail that the administrative agency has to build into the license the exact types of pesticide and amounts to be applied (of which the administrative and enforcement costs are of course enormous) the advantages of a licensing system compared to zoning seem to be limited.

Moreover, the disadvantages of licensing are well known. A detailed system of licensing is very costly to administer and the risks are high that private interests will again play an important role in the licensing system. Hence, the licensing system might result in large differences between the situation of individual farmers that cannot be explained on public interest grounds, but are probably the result of successful lobbying with the licensing administrative agency. Also the public choice literature has pointed to the market distortive effects of licensing.[46]

Summary

One can point to various instruments that could control the use of pesticides so as to reduce their accumulation in drinking water. Of course the mechanism which is finally chosen will depend upon the technical criteria that influence the accumulation problem. These may be difficult to ascertain or the information costs used to include them in the regulation may be enormous. It is these information costs which will often be crucial in the choice of instruments. For that reason it was stated that liability rules alone will not suffice to control the accumulation problem. Neither the parties in a liability setting (i.e. the farmer), nor the judge who has to fix the due care standard can be expected to have accurate information on the true social costs of pesticide use. The problem of using an accumulation tax is that the aim of the tax should be to give farmers incentives to use the right pesticide in the right quantity on a specific type of land and under specific conditions. Therefore it should be the farmer who should be taxed for the incorrect use of the pesticide and not the producer who should be taxed for the production of a certain type of pesticide. The information and enforcement costs required to apply such a tax schedule are very high. Therefore some regulatory intervention seems warranted. A simple ban will again not suffice, since, on the one hand, it neglects the substitution problem and, on the other, creates inefficiencies by prohibiting pesticides in areas where their use might be relatively harmless. A more balanced approach might be found in zoning regulations, provided that areas can be distinguished where pesticide use would lead less quickly to accumulation in drinking water than in other areas. A system of individual licensing would go too far. The additional benefits compared to a zoning system are minor and the disadvantages both with respect to administrative costs and with respect to the risks of subverting the agency are large. Hence, the accumulation problem might be best controlled through zoning regulation. Of course all this depends on the question of whether the location is indeed a decisive criterion for the accumulation problem. Moreover, liability rules should always play an additional role as a 'catch all' legal instrument that can still lead to some marginal deterrence of wrongful behaviour.

Standard-setting in a European context

Before designing a system for the regulation of pesticides we should look first at the characteristics of the regulatory framework in which this regulatory system should be constructed. The EU context has two major characteristics: the fact that an extra level of regulation is established, and the fact that this regulation can be enforced upon the lower levels of government.

Regulation of the EU level

The EU adds an extra layer of government to the already existing hierarchical structures of the Member States. As an approximation one can say, therefore, that the basic governmental structure of Member States of the EU comprises first a European level, second a Member State level, third a regional level and fourth a municipal level.[47] Each level of government has its own competencies. The EU can regulate in the field of the environment as long as it takes into account the principle of subsidiarity, which states that:

> In areas which do not fall within its exclusive competence [such as the environment], the Community shall take action, in accordance with the principle of subsidiarity, only if and in so far as the objectives of the proposed action cannot be sufficiently achieved by the Member States and can therefore, by reason of the scale or the effects of the proposed action, be better achieved by the Community.
>
> Any action by the Community shall not go beyond what is necessary to achieve the objectives of this Treaty.

The principle of subsidiarity, as laid down in Article 3B of the EC Treaty, is a general guideline for the action of the EU *inter alia*, in the field of the environment. It comprises two tests. First, it formally introduces an *efficiency test* into the EC Treaty. Matters that can be dealt with better at the EU level should be regulated by the EU; matters that are better dealt with by the Members States should be left to their competence. Secondly, it comprises the *principle of proportionality*. This principle adds to the efficiency consideration that matters which are dealt with at a European level should not regulate more than is strictly necessary. The efficiency and proportionality tests in Article 3B, if correctly used, allow for an optimal differentiation of rules with an effective distribution of competencies between the EU and the Member States.

Once the EU has adopted legislation, this legislation is binding on the national authorities, the 'lower' levels of government. As soon as a certain field or subject is harmonised, Member States lose their competence to legislate in this field, which is then an exclusive field of Community policy.[48] The remaining competence for the other layers of 'European government' in this field is the discretion that is left to the government of the Member State to implement, or transpose, the European standards into national legislation. In the case of standards for drinking water, the EU has set clear binding standards by adopting a directive which the Member States have to implement into their national legal system. A margin of discretion is, however, allowed in that 'a directive is binding, as to the result to be achieved, upon each Member State to which it is addressed, but shall leave to the national authorities the choice of form and methods' (Article 189 EC). The discretion of the Member States is therefore in choosing the way of enforcing these standards, and putting them into actual legislation, which, in the case of atrazine, Italy did by completely banning its use.[49]

Enforcement of EU legislation upon Member States

The second characteristic of the unique relation between the EU level of government and the Member States is the growing possibility of the enforcement of compliance of the European environmental standards on Member States by the EU and individuals. Individuals or environmental groups can file a complaint with the Commission in the case of poor or non-compliance by a Member State. The Commission can subsequently decide to bring infringement proceedings in the European Court of Justice under Article 169 EC.[50] Until recently the enforcement of the implementation of directives was mainly aimed at the *formal* (i.e. the text of the national legislation was in conformity with the text of the EU Directive) compliance of the Member States. The problem was however that most Member States *formally* complied with the Directives, but in practice did not enforce[51] and apply them (*practical implementation*).[52] The attitude of the Commission has, however, changed importantly. The Commission is increasingly initiating proceedings on the basis of Article 169 EC against Member States that in practice do not comply with European environmental Directives. Recent examples of this are the case of the drinking water in Verviers[53] and the case of the bathing waters at Blackpool and Southport.[54] In the case of Verviers the Commission initated 169 proceedings, and the ECJ convicted the Belgian government because the quality of the drinking water in Verviers did not comply with the quality standards as laid down in Directive 75/440/EEC on the quality of drinking water.

Until recently neither the ECJ nor the Commission had the power to enforce the compliance with the judgement of the ECJ. However, since the entry into force of the Maastricht Treaty, a new Article 171 has been introduced in the EC Treaty, stating that:

> If the Member State concerned fails to take the necessary measures to comply with the Court's judgement within the time limit laid down by the Commission, the latter may bring the case before the Court of Justice. In doing so it shall specify the amount of the lump sum or penalty payment to be paid by the Member State concerned which it considers appropriate in the circumstances
>
> If the Court of Justice finds that the Member State concerned has not complied with its judgement it may impose a lump sum or penalty payment on it

Furthermore the ECJ, in its decision in *Francovich*,[55] opened the possibility for individuals (or even environmental groups?) to hold Member States liable in national courts for the damage suffered as a result of the Member State's failure to implement the directive. In order for damages to be received, first the Directive has to grant rights to individuals, secondly it should be able to identify the content of those rights on the basis of the provisions of the directive, and thirdly there has to exist a causal link between the breach of the State's obligation to implement the EU norm and the loss and damage suffered by the injured parties.

Summary

The European context adds an extra European regulatory level to the already existing national levels of environmental legislation. This extra level makes is possible to establish within the context of the subsidiarity principle, general quality goals of European environmental standards. In addition, the new Article 171 and the rules for state liability under *Francovich* open up important new possibilities to enforce the compliance of Member States with these standards of European environmental legislation. The existence of European standards, combined with the extra 'European enforcement' makes national compliance with environmental standards more likely. In developing a regulatory framework for the prevention of the accumulation of pesticides in drinking water, these characteristics of the European context have to be taken into account.

Proposal for pesticide regulation in a European–national context

We have already given a description of the possible instruments for regulating pesticides. However, as we said before, the actual usefulness and effectiveness of a regulatory instrument also depends on the context in which it is used and the enforcement methods available. The question is therefore how, using the instruments we have examined before, the use of pesticides should be regulated in a specific EU context, taking into account the optimal differentiation of standards.

The European Union standards

We have already described some of the economic principles of standard-setting and we have indicated that as far as possible an optimal specificity of standards should be achieved, which means that differentiated standards, taking into account location-specific criteria, should be applied where possible. This poses some serious questions with respect to environmental policy within the EU. Taking into account the economic criteria for standard-setting one could argue that location-specific criteria should be taken into account and that therefore environmental standards should not be the same throughout the whole EU. The efficient standard might be relatively high in the industrial areas of the EU with a high population density and a consequent heavy load of environmental pollution, but might be more lenient in areas where the natural cleansing capacity of the environment can still absorb a certain amount of pollution. Indeed, the requirements on disposal of industrial waste-water can certainly be different for instance in the city of Antwerp than in a non-industrial area somewhere else in Europe. Obviously one has to take into account that the marginal benefit of investment in highly sophisticated environmental technology is relatively high in existing heavily polluted areas. Pollution abatement in non-industrial areas might well be possible with relatively modest technological equipment, given the much higher biodegradability of pollution in unpolluted areas. Requiring more stringent standards in Antwerp than in these other areas would therefore be efficient, since the marginal social costs of pollution differ. Requiring more stringent standards in the non-polluted, non-industrial areas or for instance the same standard as in Antwerp might even be inefficient, since it neglects a possible higher natural capacity for self-purification in the environment.

The question therefore is 'which instrument can be best used that takes into account these regional differences?' As mentioned above, the philosophy of the EU is that products such as pesticides should not be present in drinking water, as this is a fundamental resource, intended for regular daily consumption. It is therefore generally agreed that there should not be more than a specific amount of pesticides in the drinking water. The logical solution therefore seems to be to impose a general quality standard for drinking water on the Member States. This standard, however, has to be set at a level which is assumed not to cause danger to the environment and human health. At the moment the EU has already adopted this approach for the general protection of the drinking water. More stringent standards that were set in the past when compliance with Directives was to all intents and purposes optional should be revised when they are shown to be clearly inefficient.

Member States, however, have to comply with the quality standards as set. As we have seen before, the means of enforcement for both the Commission and individuals of the EU standards on Member States have increased significantly only in the last few years. At the moment only a few cases are known of the enforcement by the Commission of practical compliance with quality standards on Member States via an Article 169 procedure and of individuals using the *Francovich* construction. However, as these enforcement procedures are developing rapidly, much will be expected from them in the future.

As regards the incentives that have to be given to the producers of pesticides, it seems effective to formulate these rules on a European level, in order not to distort the free trade in pesticides. First, it should be decided, at the European level, which pesticides are under no circumstances safe to use because of their consequences for the environment and human health. Secondly, producers should be given incentives generally to develop and produce 'friendlier' pesticides, or alternative methods of crop protection. This could be done by a kind of community-wide tax system. Indeed, a location-specific tax as applied to the producer would run counter to the idea of the free movement of goods, since it would imply that products are marketed at different prices in various regions because of the different accumulation tax. Location-specific circumstances could, however, be taken into account in setting the national standards.

The national standards

As we have already seen, a simple ban on the use of a pesticide is not alone sufficient to achieve the aim of giving farmers the incentive to use pesticides under the right conditions. The regulation of the pesticides therefore needs to be differentiated according to the circumstances under which these pesticides can best be used. The properties of atrazine require regulation that takes into account the hydrological and geological conditions of the area.

As Vighi and Zanin have shown,[56] areas can be separated into four types in terms of the vulnerability of their aquifers:

(1) Areas where herbicides could be used without particular restrictions.

(2) Areas where 'leaching' herbicides could be used only in exceptional cases, and with a restriction on the amount used.

(3) Areas where only 'non-leaching' herbicides can be used and with some control of the amount used.

(4) Areas where the use of herbicides (and perhaps all pesticides) should be completely forbidden because of the vulnerability of the aquifers and their strategic value as drinking-water supplies.

To make this classification, Vighi and Zanin composed a list of necessary information, which includes *inter alia* the permeability of surface soils, the hydrogeological structure of the area, the herbicides used and their characteristics and the distribution of the crop types.[57] Although the information is not yet complete for all areas, *ad hoc* studies fill the gaps.

Using this classification of four types of vulnerability, zones can be established in which different rules for the use of pesticides may be imposed. A general regulation would be enough to set standards for the use of specific pesticides in each type of zone. The administrative and enforcement costs of such a system would be minimal. Every time a new pesticide is put on the market research has to be done on the polluting, accumulative properties of this pesticide. Once the pesticide is included in the regulation, enforcement authorities would have to take regular soil and drinking-water samples in order to determine the compliance. High penalties for non-compliance would have to be set to support the enforcement aspects of the regulatory system.

The influence of the likely public choice effects on the decisions concerning the designation of the zones and the admissibility of pesticides in these zones will be considered diminished by the enforceability of the EU quality standards on the Member States. Zoning authorities will not be able to adopt less stringent standards than practically possible because they might then be liable for a subsequent breach of the EU quality objectives.

As said before the use of zoning has already proved quite successful in planning law. There are, however, also some examples of types of zoning in environmental protection law. An example of this is the Dutch 'Besluit gebruik dierlijke meststoffen' (Regulation on the use of animal fertilisers).[58] This regulation specifies which quantities of animal fertilisers can be used on which types of soil and the periods in which the use of fertilisers is allowed. The regulation proves that zoning might be a useful instrument in pesticide regulation. Furthermore, using the instrument of zoning with fixed rules for each type of zone will not attract the considerable administrative costs incurred by giving each farmer a licence for the pesticides he or she wants to use. Also, it will allow for an optimal efficient use of pesticides, taking into account the location-specific circumstances. In the United Kingdom a parallel case exists with the control of nitrate fertilisers used in designated zones where groundwater quality can be adversely affected.

One of the advantages of a zoning system is that it can be used in combination with other types of regulatory instrument. Instead of imposing detailed standards on the use of the pesticides in the specific zones, the national authorities might choose economic instruments for giving the farmers the right incentives. An example might be putting a higher tax, under specific circumstances, on the use of more polluting pesticides. The problem might be, however, that this could entail high administrative and enforcement costs.

Another important advantage of the zoning system is that in regulating the use of pesticides with reference to the hydrogeological situation, the users of pesticides will automatically create a market for pesticides that are more fit to be used within a specific area. As the use of pesticides in a certain zone is permitted on the basis of their accumulation or pollution value, farmers will try to find pesticides which pollute less under those circumstances. An open pesticide market will automatically try to develop new products to satisfy the newly created demands. The market itself will give producers the incentive to create a broader spectrum of pesticides for specific hydrogeological conditions.

Conclusion

In this chapter we have tried to establish a framework for the optimal regulation of the use of pesticides, specifically atrazine, on the basis of their property of accumulating in drinking water. We concluded from a review of previous research that the current regulatory approach to the accumulation of pesticides in drinking water is not working properly, in that it only transfers the environmental problems to other areas or other types of pesticide. We found that the aim of the regulation of pesticides should be twofold. Firstly, and most importantly, incentives should be given to producers of pesticides to produce 'less polluting' alternatives for atrazine. Secondly, we found that farmers should be given incentives to use certain pesticides under the right conditions.

We discussed the law and economics of standard setting and optimal differentiation in order to point out that the current system is inefficient and in order to find criteria for efficient regulatory methods. With reference to the current institutional framework of regulation and enforcement in a European and national context we tried to take the advantages of this framework into account in proposing a regulation of pesticides.

We believe that the most efficient way of regulating the use of pesticides in order to protect the drinking water is by setting different rules on two different levels. First of all, quality standards have to be set at the European level. These quality standards should be enforceable upon the Member States by both the Commission and individuals who suffer damage because the Member States do not meet the Union standards. Member States can most efficiently reach these goals by using a system of zoning of regions where pesticides are frequently used. In each type of zone rules should be set that ensure the correct use of pesticides with regard to the characteristics of the zone. The system of zoning allows for an optimal use of pesticides, leading to an acceptably low amount of accumulation in the drinking water. Of course these zoning rules have to be enforced upon the farmers by the State or regional authorities. The 'invisible hand' of the open market for pesticides will then automatically create incentives for pesticide producers to produce less-accumulating pesticides. This system could be supplemented by European legislation giving the producers of pesticides incentives to produce 'cleaner' chemicals, possibly using a tax system, making the framework of optimal regulation of the market and the use of pesticides complete.

We feel that designing and implementing a regulatory framework as described above will lead to an efficient, optimally differentiated regulation of the use of pesticides, in the context of the goal of preventing the accumulation of pesticides in drinking water.

Notes

1 This paper was made possible by financial support from the European Science Foundation, which we gratefully acknowledge.
2 Faure, 1994.
3 Directive 80/778/EEC of 15 July 1980 relating to the quality of water intended for human consumption.
4 Vighi and Zanin, 1994.
5 Vighi and Zanin 1994, p. 112.
6 See Faure, 1994, pp. 57–8.
7 Vighi *et al.*, Chapter 4, this volume.
8 Atrazine does indeed seem to have significant, previously unknown effects on human health, causing *inter alia*, breast-cancer, *ENDS Report* **241**, February 1995.
9 Judgement of the European Court of Justice (ECJ) of 19 November 1991, *Andrea Francovich and Danila Bonifaci and others* v. *Italian Republic*, Joined cases C-6/90 and C-9/90, ECR [1991] page i-5357.
10 Judgement of the ECJ 9 July 1992, *Commission of the European Communities* v. *Kingdom of Belgium*, Case C-2/90, Reports of cases 1992 pages i-4431, Judgement of the ECJ of 14 July 1993, *Commission of the EC* v. *United Kingdom of Great Britain and Northern Ireland*, Case C56/90, Reports of Cases 1993, page i-4109.
11 The New Article 171 of the EC Treaty, introduced by the Maastricht Treaty on the EU.
12 See pp. 266–9.
13 A good example of the problems caused by *inter alia* high EC standards in the drinking water Directive is also the case of *Cambridge Water Company LTD.* v *Eastern Counties Leather PLC.*, House of Lords, 9 December 1993, [1994] *Environmental Law Reports*, p. 105–33.
14 Zanin, *et al.* [1993]. Example taken from Söderqvist, 1994.
15 This point concerning market distortions is further developed in Faure, 1994, pp. 77–80.
16 This has indeed happened in Italy, see Vighi and Zanin, 1994.
17 See Swanson, Chapter 9, this volume.
18 See p. 250–1: the precise level of admissible atrazine is, however, not a point of discussion here.
19 Vighi and Zanin, pp. 124–5, and Zanin *et al.* (1993).

20 Vighi and Zanin, 1994, p. 125.

21 Vighi and Zanin, 1994.

22 A third possibility would be to consider cleaning the polluted groundwater if it is used, for example, as drinking water. Within the scope of this paper we do not examine this further.

23 Risk neutrality is assumed here. In the case of risk aversion higher investments in environmental technology might be efficient since they can remove disutility of risk from risk-averse persons.

24 Söderqvist, 1994, 151–71.

25 See Ogus, 1994*a*; 1994*b*, pp. 174–9.

26 Ehrlich and Posner, 1974, p. 257; Ogus, 1981, pp. 210–25; Ogus, 1981, pp. 168–71.

27 'Capturing' refers to a subversive pressure being put on the agency by the regulated.

28 See Ogus, 1994*a*.

29 Posner, 1986, p. 513.

30 See Ogus, 1992, pp. 411–21; Mackaay, 1982; Schwarz and Wilde, 1979, pp. 630–82; Stigler, 1961, p. 213.

31 Directive 82/176/EEC of 22 March 1982.

32 Directive 89/428/EEC of 21 June 1989.

33 Tollison, 1982, p. 590.

34 See Shavell, 1984; and see the discussion of this literature in Faure, 1994, pp. 42–4.

35 Swanson, Chapter 9, this volume.

36 Dewees, 1992*a, b*.

37 Rose-Ackerman, 1992*a*; 1992*b*, pp. 118–31.

38 Mason, Chapter 9, this volume.

39 Posner, 1986, pp. 351–8.

40 They are therefore not the 'cheapest cost avoider'.

41 Posner, 1986, p. 354.

42 Shavell, 1984, pp. 363–4.

43 Swanson, Chapter 9, this volume.

44 In the law and economics literature zoning is discussed as a solution to the problem of conflicting land uses (Posner, 1986, pp. 60–1) and as a response to market failure (Cooter, and Ulen, 1988, pp. 205–9). For an example of planning controls around hazardous installations see Rocard and Smets, 1992, 468–84.

45 See Cooter and Ulen, 1988, pp. 208–9.

46 See Moore, 1961.

47 Of course depending on the constitutional structure of a Member State there can be more levels, such as the German Bundesländer and the Spanish Communidades Autónomas.

48 Case 22/70, *Commission* v. *Council (Re European Road Transport Agreement)*, [1971] ECR 263.

49 A clear description of the directive and its implementation measures in Italy has already been given in Faure, 1994.

50 See Harlow (1992, p. 343), who gives the example of the British environmental groups using the complaint procedure to force the UK to comply with the Bathing Waters Directive.

51 The formal implementation might take place because of its symbolic value. One can often note that after a symbolic implementation no resources are allocated to monitoring and enforcement activities; compare Han, 1990, p. 35.

52 Tenth report of the Commission on the application of Community Law, COM (93) 320 Final of 28 April 1993, pp. 101–2.

53 ECJ of 5 July 1990, Case C-42/89, *Commission of the European Communities* v. *Kingdom of Belgium*, ECR [1990], page i-2821.

54 ECJ decison of 14 July 1993, *Commission* v. *United Kingdom*, Case C-56/90, unreported.

55 Cases 6&9/90, *Francovich and Bonifaci* v. *Italian Republic*, ECR [1991], page i-5357.

56 Vighi, and Zanin, 1994, pp. 124–5.

57 Vighi and Zanin, 1994.

58 *Staatsblad* 1987, 114, changed in *Staatsblad* 1989, 501 and *Staatsblad* 1990, 99.

References

Cooter, R. and Ulen, T. (1988). *Law and Economics*. Scott, Foresman & Co., Glenview, IL.

Dewees, D. (1992*a*). The comparative efficacy of tort law and regulation for environmental protection. *Geneva Papers on Risk and Insurance*, **65**, 446–67.

Dewees, D. (1992*b*) Tort law and the deterrence of environmental pollution. In *Innovation in Environmental Policy, Economic and Legal Aspects of Recent Developments in Environmental Enforcement and Liability*, ed. T. H. Tietenberg, pp. 139–64. Elger, Brookfield.

Ehrlich, I. and Posner, R. (1974). An economic analysis of legal rule-making. *Journal of Legal Studies*, **3**, 257–86.

Faure, M. (1994). The EC Directive on drinking water: institutional aspects. In *Environmental Toxicology, Economics and Institutions: The Atrazine Case Study*, ed. L. Bergman and D. M. Pugh, pp. 39–87. Kluwer Academic Publisher, Dordrecht.

Hahn, R. (1990). The political economy of environmental regulation: towards a unifying framework, *Public Choice*, **65**, 21–47.

Harlow, C. (1992). *A community of interests? Making the most of European Law.* Modern Law review 55, 331–50.

Mackaay, E. (1982) *Economics of Information and the Law*, Martinus Nijhoff, Boston, MA.

Moore, T. (1961) The purpose of licensing. *Journal of Law and Economics*, **4**, 93–117.

Ogus, A. I. (1981) Quantitative rules and judicial decision-making. In *The Economic Approach to Law*, ed. P. Burrows and C. Veliyanowski, pp. 210–25. Butterworths, London.

Ogus, A. I. (1992) Information, error costs and regulation. *International Review of Law and Economics*, 12, 411–21.

Ogus, A. I. (1994*a*) *Standard setting for environmental protection: principles and processes.* In *Environmental Standards in the European Union in an Interdisciplinary Framework,*, ed. M. Faure, J. Vervaele and A. Weale. Maklu, Antwerp.

Ogus, A. I. (1994*b*) *Regulation Legal Form and Economic Theory*, Clarendon Press, Oxford.

Posner, R. (1986) *Economic Analysis of Law*, third edition. Little, Brown and Co., Boston, MA.

Rocard, Ph. and Smets, H. (1992) A socio-economic analysis of controls on land use around hazardous installations. *Geneva Papers on Risk and Insurance*, 65, 468–84.

Rose-Ackerman, S. (1992*a*) Environmental liability law. In *Innovation in Environmental Policy, Economic and Legal Aspects of Recent Developments in Environmental Enforcement and Liability*, ed. T. H. Tietenberg, pp. 223–43. Elgar, Brookfield.

Rose-Acerman, S. (1992) *Rethinking the Progressive Agenda: The Reform of the American Regulatory State.* Free Press. New York.

Schwarz, A. and Wilde, L. (1979) Intervening in markets on the basis of imperfect information: a legal and economic analysis. *University of Pennsylvania Law Review*, 127, 630–82.

Shavell, S. (1984). Liability for harm versus regulation of safety. *Journal of Legal Studies*, 13, 357–74.

Söderqvist, T. (1994). The costs of meeting a drinking water quality standard: the case of atrazine in Italy. In *Environmental Toxicology, Economics and Institutions: The Atrazine Case Study*, ed. L. Bergman and D. M. Pugh, pp. 151–71. Kluwer Academic Publishers, Dordrecht.

Stigler, G. (1961). The economics of information. *Journal of Political Economy*, 69, 213–25.

Tollison, R. (1982). Rent seeking: a survey. *Kyklos*, 35, 575–602.

Vighi, M. and Zanin, G. (1994). Agronomic and ecotoxicological aspects of herbicide contamination of groundwater in Italy. In *Environmental Toxicology, Economics and Institutions*, ed. L. Bergman and D. M. Pugh, pp. 111–25. Kluwer Academic Publishers, Dordrecht.

Zanin, G., Borin, M., Altissimo, L. and Calamari, D. (1993). Simulation of herbicide contamination of the aquifer north of Vicenza (north-east Italy). *Chemosphere*, 26, 929–40.